MEI STRUCTURED MATHEMATICS

SECOND EDITION

# Pure Mathematics 3

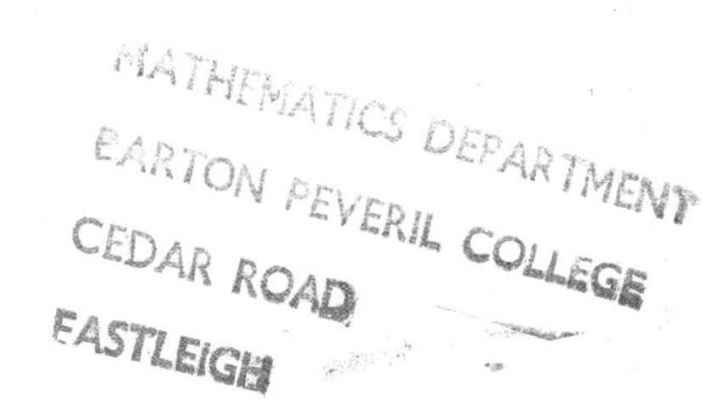


Catherine Berry
Val Hanrahan
Roger Porkess

Series Editor: Roger Porkess

Hodder & Stoughton
A MEMBER OF THE HODDER HEADLINE GROUP

## Acknowledgements

We are grateful to the following companies, institutions and individuals who have given permission to reproduce photographs in this book. Every effort has been made to trace and acknowledge ownership of copyright. The publishers will be glad to make suitable arrangements with any copyright holders whom it has not been possible to contact.

Guinness Records (page 1); Mark Ferguson/Life File (page 27); D. Boone/CORBIS (page 27); Phil Cole/Allsport (page 34); Robert Harding (page 38 and 152); Marc Garanger/CORBIS (page 48); Topham Picturepoint (page 60 and 162); Jeremy Hoare/Life File (page 98); Robert Whistler/Life File (page 123); Hodder Picture Library (page 177); Emma Lee/Life File (page 179).



Orders: please contact Bookpoint Ltd, 130 Milton Park, Abingdon, Oxon OX14 4SB.
Telephone: (44) 01235 827720, Fax: (44) 01235 400454. Lines are open from 9.00-6.00, Monday to Saturday, with a 24 hour message answering service.
Email address: orders@bookpoint.co.uk

*British Library Cataloguing in Publication Data*
A catalogue record for this title is available from the The British Library

ISBN 0 340 77 1968

First published 1995
Second edition published 2000
Impression number 10 9 8 7 6 5 4 3
Year 2005 2004 2003 2002 2001



Typeset by Pantek Arts Ltd, Maidstone, Kent.
Printed in Great Britain for Hodder & Stoughton Educational, a division of Hodder Headline Plc, 338 Euston Road, London NW1 3BH by Martins the Printers Ltd, Berwick upon Tweed.

# MEI Structured Mathematics

Mathematics is not only a beautiful and exciting subject in its own right but also one that underpins many other branches of learning. It is consequently fundamental to the success of a modern economy.

MEI Structured Mathematics is designed to increase substantially the number of people taking the subject post-GCSE, by making it accessible, interesting and relevant to a wide range of students.

It is a credit accumulation scheme based on 45 hour modules which may be taken individually or aggregated to give Advanced Subsidiary (AS) and Advanced GCE (A Level) qualifications in Mathematics, Further Mathematics and related subjects (like Statistics). The modules may also be used to obtain credit towards other types of qualification.

The course is examined by OCR (previously the Oxford and Cambridge Schools Examination Board) with examinations held in January and June each year.

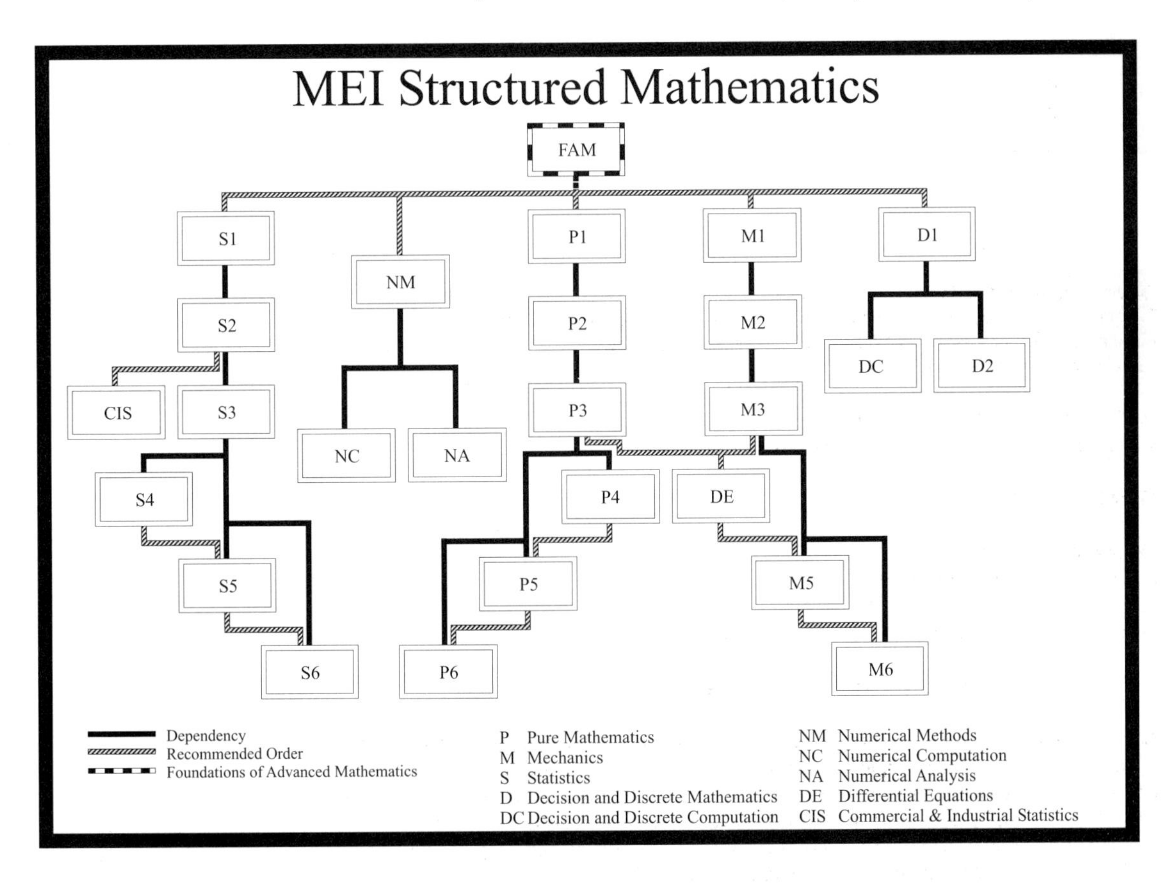


This is one of the series of books written to support the course. Its position within the whole scheme can be seen in the diagram above.

*Mathematics in Education and Industry is a curriculum development body which aims to promote the links between Education and Industry in Mathematics at secondary level, and to produce relevant examination and teaching syllabuses and support material. Since its foundation in the 1960s, MEI has provided syllabuses for GCSE (or O Level), Additional Mathematics and A Level.*

*For more information about MEI Structured Mathematics or other syllabuses and materials, write to MEI Office, Albion House, Market Place, Westbury, Wiltshire, BA13 3DE.*

# Introduction

This is the third in the series of books written to support the Pure Mathematics modules of MEI Structured Mathematics but you may also use them for an independent course in the subject. Together *Pure Mathematics 1, 2* and *3* cover the subject criteria for A Level Mathematics.

Throughout the series the emphasis is on understanding rather than on mere routine calculations, but the various exercises do provide plenty of scope for practising techniques.

In the first two chapters of this book, you meet further techniques in algebra and trigonometry, many of which are then used in the later chapters. In Chapter 3 the range of functions which you are able to differentiate and integrate is extended (including trigonometric functions) and in Chapter 6 the work on calculus is taken forward to differential equations. The other two chapters cover parametric co-ordinates and vector geometry in two and three dimensions.

In a few places the text goes a little beyond the limits of the MEI subject criteria in order to develop interesting applications or useful techniques. These are indicated by a thin line down the right-hand side of the page.

This is the second edition of this book, revised in line with the requirements of the new subject criteria being first taught in September 2000.

Thanks are due to Val Hanrahan for her work in preparing the new edition. We would also like to thank the many people who have helped this book along the way, reading the text and contributing their ideas. We also thank the various examination boards who have given permission for their past questions to be included in the exercises.

Catherine Berry, Val Hanrahan and Roger Porkess

# Contents

# 1 Algebra

**At the age of twenty-one he wrote a treatise upon the Binomial Theorem. ... On the strength of it, he won the Mathematical Chair at one of our smaller Universities.**

*Sherlock Holmes on Professor Moriarty*
*'The Final Problem' by Sir Arthur Conan Doyle*

How would you find $\sqrt{101}$ correct to 3 decimal places, without using a calculator?

Many people are able to develop a very high degree of skill in mental arithmetic, particularly those, such as bookmakers, whose work calls for quick reckoning. There are also those who have quite exceptional innate skills. M. Hari Prasad, pictured right, is famous for his mathematical speed; on one occasion he found the square root of a six digit number in just 1 minute 3.8 seconds.

While most mathematicians do not have M. Hari Prasad's high level of talent with numbers, they do acquire a sense of when something looks right or wrong. This often involves finding approximate values of numbers, such as $\sqrt{101}$, using methods that are based on series expansions, and these are the subject of the first part of this chapter.

## INVESTIGATION

Using your calculator, write down the values of $\sqrt{1.02}$, $\sqrt{1.04}$, $\sqrt{1.06}$, ..., giving your answers correct to 2 decimal places.

What do you notice?

Use your results to complete the following, giving the value of the constant $k$.

$$\sqrt{1.02} = (1 + 0.02)^{\frac{1}{2}} \approx 1 + 0.02k$$

$$\sqrt{1.04} = (1 + 0.04)^{\frac{1}{2}} \approx 1 + 0.04k$$

What is the largest value of $x$ such that $\sqrt{1 + x} \approx 1 + kx$ is true for the same value of $k$?

## The general binomial expansion

You have already met the binomial expansion in the form

$$(1+x)^n = 1 + \binom{n}{1}x + \binom{n}{2}x^2 + \binom{n}{3}x^3 + \ldots + \binom{n}{r}x^r + \ldots$$

which holds when $n$ is any positive integer (or 0), that is $n \in \mathbb{N}$.

This may also be written as

$$(1+x)^n = 1 + nx + \frac{n(n-1)}{2!}x^2 + \frac{n(n-1)(n-2)}{3!}x^3 + \ldots$$
$$+ \frac{n(n-1)(n-2)\ldots(n-r+1)}{r!}x^r + \ldots$$

which, being the same expansion as above, also holds when $n \in \mathbb{N}$.

The general binomial theorem states that this second form, that is

$$(1+x)^n = 1 + nx + \frac{n(n-1)}{2!}x^2 + \frac{n(n-1)(n-2)}{3!}x^3 + \ldots$$
$$+ \frac{n(n-1)(n-2)\ldots(n-r+1)}{r!}x^r + \ldots$$

is true when ***n* is any real number**, but there are two important differences to note when $n \notin \mathbb{N}$:

- the series is infinite (or non-terminating);
- the expansion of $(1+x)^n$ is valid only if $|x| < 1$.

Proving this result is well beyond the scope of an A-level course but you can assume that it is true.

Consider now the coefficients in the binomial expansion:

$$1 \quad n \quad \frac{n(n-1)}{2!} \quad \frac{n(n-1)(n-2)}{3!} \quad \frac{n(n-1)(n-2)(n-3)}{4!} \quad \ldots$$

| | | | | | | | |
|---|---|---|---|---|---|---|---|
| When $n = 0$, we get | 1 | 0 | 0 | 0 | 0 | … | (infinitely many zeros) |
| $n = 1$ | 1 | 1 | 0 | 0 | 0 | … | ditto |
| $n = 2$ | 1 | 2 | 1 | 0 | 0 | … | ditto |
| $n = 3$ | 1 | 3 | 3 | 1 | 0 | … | ditto |
| $n = 4$ | 1 | 4 | 6 | 4 | 1 | … | ditto |

so that, for example

$$(1+x)^2 = 1 + 2x + x^2 + 0x^3 + 0x^4 + 0x^5 + \dots$$

$$(1+x)^3 = 1 + 3x + 3x^2 + x^3 + 0x^4 + 0x^5 + \dots$$

$$(1+x)^4 = 1 + 4x + 6x^2 + 4x^3 + x^4 + 0x^5 + \dots$$

Of course, it is usual to discard all the zeros and write these binomial coefficients in the familiar form of Pascal's triangle:

```
            1
        1       1
    1       2       1
1       3       3       1
1   4       6       4       1
```

and the expansions as

$$(1+x)^2 = 1 + 2x + x^2$$

$$(1+x)^3 = 1 + 3x + 3x^2 + x^3$$

$$(1+x)^4 = 1 + 4x + 6x^2 + 4x^3 + x^4$$

However, for other values of $n$ (where $n \notin \mathbb{N}$) there are no zeros in the row of binomial coefficients and so we obtain an infinite sequence of non-zero terms. For example:

$n = -3$ gives $1 \quad -3 \quad \frac{(-3)(-4)}{2!} \quad \frac{(-3)(-4)(-5)}{3!} \quad \frac{(-3)(-4)(-5)(-6)}{4!} \dots$

that is $1 \quad -3 \quad 6 \quad -10 \quad 15 \dots$

$n = \frac{1}{2}$ gives $1 \quad \frac{1}{2} \quad \frac{(\frac{1}{2})(-\frac{1}{2})}{2!} \quad \frac{(\frac{1}{2})(-\frac{1}{2})(-\frac{3}{2})}{3!} \quad \frac{(\frac{1}{2})(-\frac{1}{2})(-\frac{3}{2})(-\frac{5}{2})}{4!} \dots$

that is $1 \quad \frac{1}{2} \quad -\frac{1}{8} \quad \frac{1}{16} \quad -\frac{5}{128} \dots$

so that $\quad (1+x)^{-3} = 1 - 3x + 6x^2 - 10x^3 + 15x^4 + \dots$

and $\quad (1+x)^{\frac{1}{2}} = 1 + \frac{1}{2}x - \frac{1}{8}x^2 + \frac{1}{16}x^3 - \frac{5}{128}x^4 + \dots$

---

But remember: these two expansions are valid only if $|x| < 1$.

---

These examples confirm that there will be an infinite sequence of non-zero coefficients when $n \notin \mathbb{N}$. You can also see that, after a certain stage, the remaining terms of the sequence will alternate in sign.

In the investigation at the beginning of this chapter you showed that

$$\sqrt{1+x} \approx 1 + \tfrac{1}{2}x$$

is a good approximation for small values of $x$. Notice that these are the first two terms of the binomial expansion for $n = \frac{1}{2}$. If you include the third term, the approximation is

$$\sqrt{1+x} \approx 1 + \tfrac{1}{2}x - \tfrac{1}{8}x^2$$

Take $\quad y = 1 + \frac{1}{2}x$

$\qquad\quad y = 1 + \frac{1}{2}x - \frac{1}{8}x^2$

and $\quad y = \sqrt{1+x}$

They are shown in the graph in figure 1.1 for values of $x$ between –1 and 1.

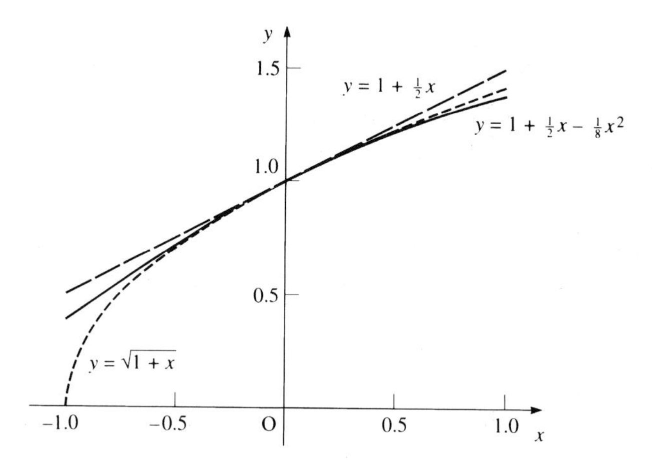


**Figure 1.1**

## INVESTIGATION

For $n = \frac{1}{2}$ the first three terms of the binomial expansion are $1 + \frac{1}{2}x - \frac{1}{8}x^2$. Use your calculator to verify the approximate result

$$\sqrt{1+x} \approx 1 + \tfrac{1}{2}x - \tfrac{1}{8}x^2$$

for 'small' values of $x$. What values of $x$ can be considered as 'small' if you want the result to be correct to 2 decimal places?

Now take $n = -3$. Using the coefficients found earlier suggests the approximate result

$$(1+x)^{-3} \approx 1 - 3x + 6x^2.$$

Comment on values of $x$ for which this approximation is valid.

When $|x| < 1$, the magnitudes of $x^2, x^3, x^4, x^5, \ldots$ form a decreasing geometric sequence. In this case, the binomial expansion converges (just as a geometric series converges for $-1 < r < 1$, where $r$ is the common ratio) and has a sum to infinity.

ACTIVITY

Compare the geometric series $1 - x + x^2 - x^3 + \dots$ with the series obtained by putting $n = -1$ in the binomial expansion. What do you notice?

To summarise: when $n$ is not a positive integer or 0, the binomial expansion of $(1 + x)^n$ becomes an infinite series, and is only valid when some restriction is placed on the values of $x$.

The binomial theorem states that for any value of $n$:

$$(1 + x)^n = 1 + nx + \frac{n(n-1)}{2!}x^2 + \frac{n(n-1)(n-2)}{3!}x^3 + \dots$$

where

- if $n \in \mathbb{N}$, $x$ may take any value;
- if $n \notin \mathbb{N}$, $|x| < 1$.

***Note***

The full statement is the binomial *theorem*, and the right-hand side is referred to as the binomial *expansion*.

**EXAMPLE 1.1**

Expand $(1 - x)^{-2}$ as a series of ascending powers of $x$ up to and including the term in $x^3$, stating the set of values of $x$ for which the expansion is valid.

**SOLUTION**

$$(1 + x)^n = 1 + nx + \frac{n(n-1)}{2!}x^2 + \frac{n(n-1)(n-2)}{3!}x^3 + \dots$$

Replacing $n$ by $-2$, and $x$ by $(-x)$ gives

$$(1 + (-x))^{-2} = 1 + (-2)(-x) + \frac{(-2)(-3)}{2!}(-x)^2$$

$$+ \frac{(-2)(-3)(-4)}{3!}(-x)^3 + \dots \quad \text{when } |-x| < 1$$

It is important to put brackets round the term $-x$, since, for example, $(-x)^2$ is not the same as $-x^2$

which leads to

$$(1 - x)^{-2} \approx 1 + 2x + 3x^2 + 4x^3 \qquad \text{when } |x| < 1.$$

***Note***

In this example the coefficients of the powers of $x$ form a recognisable sequence, and it would be possible to write down a general term in the expansion. The coefficient is always one more than the power, so the $r$th term would be $rx^{r-1}$. Using sigma notation, the infinite series could be written as

$$\sum_{r=1}^{\infty} rx^{r-1}.$$

**EXAMPLE 1.2**

Find a quadratic approximation for $\dfrac{1}{\sqrt{1+2t}}$ and state for which values of $t$ the expansion is valid.

**SOLUTION**

$$\frac{1}{\sqrt{1+2t}} = \frac{1}{(1+2t)^{\frac{1}{2}}} = (1+2t)^{-\frac{1}{2}}$$

The binomial theorem states that

$$(1+x)^n = 1 + nx + \frac{n(n-1)}{2!}x^2 + \frac{n(n-1)(n-2)}{3!}x^3 + \ldots$$

Remember to put brackets round the term $2t$, since $(2t)^2$ is not the same as $2t^2$

Replacing $n$ by $-\frac{1}{2}$ and $x$ by $2t$ gives

$$(1+2t)^{-\frac{1}{2}} = 1 + (-\tfrac{1}{2})(2t) + \frac{(-\frac{1}{2})(-\frac{3}{2})}{2!}(2t)^2 + \ldots \quad \text{when } |2t| < 1$$

$$\Rightarrow \quad (1+2t)^{-\frac{1}{2}} \approx 1 - t + \tfrac{3}{2}t^2 \quad \text{when } |t| < \tfrac{1}{2}.$$

**INVESTIGATION**

Example 1.1 showed how using the binomial expansion for $(1-x)^{-2}$ gave a sequence of coefficients of powers of $x$ which was easily recognisable, so that the particular binomial expansion could be written using the sigma notation. Investigate whether a recognisable pattern is formed by the coefficients in the expansions of $(1-x)^n$ for any other negative integers $n$.

The equivalent binomial expansion of $(a+x)^n$ when $n$ is not a positive integer is rather unwieldy. It is easier to start by taking $a$ outside the brackets:

$$(a+x)^n = a^n\left(1+\frac{x}{a}\right)^n$$

The first entry inside the bracket is now 1 and so the first few terms of the expansion are

$$(a+x)^n = a^n\left[1 + n\left(\frac{x}{a}\right) + \frac{n(n-1)}{2!}\left(\frac{x}{a}\right)^2 + \frac{n(n-1)(n-2)}{3!}\left(\frac{x}{a}\right)^3 + \ldots\right]$$

$$\text{for } \left|\frac{x}{a}\right| < 1.$$

***Note***

Since the bracket is raised to the power $n$, any quantity you take out must be raised to the power $n$ too, as in the following example.

**EXAMPLE 1.3**

Expand $(2 + x)^{-3}$ as a series of ascending powers of $x$ up to and including the term in $x^2$, stating the values of $x$ for which the expansion is valid.

**SOLUTION**

$$(2 + x)^{-3} = \frac{1}{(2 + x)^3}$$

$$= \frac{1}{2^3\left(1 + \frac{x}{2}\right)^3}$$

$$= \frac{1}{8}\left(1 + \frac{x}{2}\right)^{-3}$$

Notice that this is the same as $2^{-3}\left(1 + \frac{x}{2}\right)^{-3}$

Take the binomial expansion

$$(1 + x)^n = 1 + nx + \frac{n(n-1)}{2!}x^2 + \frac{n(n-1)(n-2)}{3!}x^3 + \ldots$$

and replace $n$ by $-3$ and $x$ by $\frac{x}{2}$ to give

$$\frac{1}{8}\left(1 + \frac{x}{2}\right)^{-3} = \frac{1}{8}\left[1 + (-3)\left(\frac{x}{2}\right) + \frac{(-3)(-4)}{2!}\left(\frac{x}{2}\right)^2 + \ldots\right] \quad \text{when } \left|\frac{x}{2}\right| < 1$$

$$\approx \frac{1}{8} - \frac{3x}{16} + \frac{3x^2}{16} \quad \text{when } |x| < 2.$$

---

? The chapter began by asking how you would find $\sqrt{101}$ to 3 decimal places without using a calculator. How would you find it?

---

**EXAMPLE 1.4**

Find a quadratic approximation for $\frac{(2 + x)}{(1 - x^2)}$, stating the values of $x$ for which the expansion is valid.

**SOLUTION**

$$\frac{(2 + x)}{(1 - x^2)} = (2 + x)(1 - x^2)^{-1}$$

Take the binomial expansion

$$(1 + x)^n = 1 + nx + \frac{n(n-1)}{2!}x^2 + \frac{n(n-1)(n-2)}{3!}x^3 + \ldots$$

and replace $n$ by $-1$ and $x$ by $(-x^2)$ to give

$$(1 + (-x^2))^{-1} = 1 + (-1)(-x^2) + \frac{(-1)(-2)(-x^2)^2}{2!} + \ldots \quad \text{when } |-x^2| < 1$$

$$(1 - x^2)^{-1} = 1 + x^2 + \ldots \quad \text{when } |x^2| < 1, \text{ i.e. when } |x| < 1.$$

Multiply both sides by $(2 + x)$ to obtain $(2 + x)(1 - x^2)^{-1}$:

$$(2 + x)(1 - x^2)^{-1} \approx (2 + x)(1 + x^2)$$

$$\approx 2 + x + 2x^2 \quad \text{when } |x| < 1.$$

The term in $x^3$ has been omitted because the question asked for a quadratic approximation

Sometimes two or more binomial expansions may be used together. If these impose different restrictions on the values of $x$, you need to decide which is the strictest.

**EXAMPLE 1.5**

Find $a$ and $b$ such that

$$\frac{1}{(1 - 2x)(1 + 3x)} \approx a + bx$$

and state the values of $x$ for which the expansion is valid.

**SOLUTION**

$$\frac{1}{(1 - 2x)(1 + 3x)} = (1 - 2x)^{-1}(1 + 3x)^{-1}$$

Using the binomial expansion:

$$(1 - 2x)^{-1} \approx 1 + (-1)(-2x) \quad \text{for } |-2x| < 1$$

and $$(1 + 3x)^{-1} \approx 1 + (-1)(3x) \quad \text{for } |3x| < 1$$

$$\Rightarrow \quad (1 - 2x)^{-1}(1 + 3x)^{-1} \approx (1 + 2x)(1 - 3x)$$

$$\approx 1 - x \quad \text{(ignoring higher powers of } x\text{)}$$

giving $a = 1$ and $b = -1$.

For the result to be valid, both $|2x| < 1$ and $|3x| < 1$ need to be satisfied.

$$|2x| < 1 \quad \Rightarrow \quad -\tfrac{1}{2} < x < \tfrac{1}{2}$$

and $$|3x| < 1 \quad \Rightarrow \quad -\tfrac{1}{3} < x < \tfrac{1}{3}.$$

Both of these restrictions are satisfied if $-\frac{1}{3} < x < \frac{1}{3}$. This is the stricter restriction.

*Note*

The binomial expansion may also be used when the first term is the variable. For example:

$(x+2)^{-1}$ may be written as $(2+x)^{-1} = 2^{-1}\left(1+\frac{x}{2}\right)^{-1}$

and $(2x-1)^{-3} = [(-1)(1-2x)]^{-3}$

$= (-1)^{-3}(1-2x)^{-3}$

$= -(1-2x)^{-3}$.

---

**?** What happens when you try to rearrange $\sqrt{x-1}$ so that the binomial expansion can be used?

---

**EXERCISE 1A**

*For each of the functions in questions* 1 *to* 12:

**(i)** write down the first three non-zero terms in their expansions as a series of ascending powers of $x$;

**(ii)** state the values of $x$ for which the expansion is valid;

**(iii)** substitute $x = 0.1$ in both the function and its expansion and calculate the relative error, where

$$\text{relative error} = \frac{\text{absolute error} \times 100}{\text{true value}}\%.$$

**(iv)** If you have access to a graphics calculator or suitable computer package, draw the graphs of each function and the first three terms of its binomial expansion on the same axes. In each case, notice how the graphs illustrate the need for some restriction on the values of $x$.

**1** $(1+x)^{-2}$ **2** $\frac{1}{1+2x}$ **3** $\sqrt{1-x^2}$

**4** $\frac{1+2x}{1-2x}$ **5** $(3+x)^{-1}$ **6** $(1-x)\sqrt{4+x}$

**7** $\frac{x+2}{x-3}$ **8** $\frac{1}{\sqrt{3x+4}}$ **9** $\frac{1+2x}{(2x-1)^2}$

**10** $\frac{1+x^2}{1-x^2}$ **11** $\sqrt[3]{1+2x^2}$ **12** $\frac{1}{(1+2x)(1+x)}$

**13 (i)** Write down the expansion of $(1+x)^3$.

**(ii)** Find the first four terms in the expansion of $(1-x)^{-4}$ in ascending powers of $x$. For what values of $x$ is this expansion valid?

**(iii)** When the expansion is valid

$$\frac{(1+x)^3}{(1-x)^4} = 1 + 7x + ax^2 + bx^3 + \dots$$

Find the values of $a$ and $b$. [MEI]

**14 (i)** Write down the expansion of $(2 - x)^4$.

**(ii)** Find the first four terms in the expansion of $(1 + 2x)^{-3}$ in ascending powers of $x$. For what range of values of $x$ is this expansion valid?

**(iii)** When the expansion is valid

$$\frac{(2-x)^4}{(1+2x)^3} = 16 + ax + bx^2 + \ldots$$

Find the values of $a$ and $b$.

[MEI]

**15** Write down the expansions of the following expressions in ascending powers of $x$, as far as the term containing $x^3$. In each case state the values of $x$ for which the expansion is valid.

**(i)** $(1 - x)^{-1}$

**(ii)** $(1 + 2x)^{-2}$

**(iii)** $\dfrac{1}{(1-x)(1+2x)^2}$

[MEI]

**16 (i)** Show that $\dfrac{1}{\sqrt{4-x}} = \dfrac{1}{2}\left(1 - \dfrac{x}{4}\right)^{-\frac{1}{2}}$.

**(ii)** Write down the first three terms in the binomial expansion of $\left(1 - \dfrac{x}{4}\right)^{-\frac{1}{2}}$ in ascending powers of $x$, stating the range of values of $x$ for which this expansion is valid.

**(iii)** Find the first three terms in the expansion of $\dfrac{2(1+x)}{\sqrt{4-x}}$ in ascending powers of $x$, for small values of $x$.

[MEI]

**17 (i)** Expand $(1 + y)^{-1}$, where $-1 < y < 1$, as a series in powers of $y$, giving the first four terms.

**(ii)** Hence find the first four terms of the expansion of $\left(1 + \dfrac{2}{x}\right)^{-1}$ where $-1 < \dfrac{2}{x} < 1$.

**(iii)** Show that $\left(1 + \dfrac{2}{x}\right)^{-1} = \dfrac{x}{x+2} = \dfrac{x}{2}\left(1 + \dfrac{x}{2}\right)^{-1}$.

**(iv)** Find the first four terms of the expansion of $\dfrac{x}{2}\left(1 + \dfrac{x}{2}\right)^{-1}$ where $-1 < \dfrac{x}{2} < 1$.

**(v)** State the conditions on $x$ under which your expansions for $\left(1 + \dfrac{2}{x}\right)^{-1}$ and $\dfrac{x}{2}\left(1 + \dfrac{x}{2}\right)^{-1}$ are valid and explain briefly why your expansions are different.

[MEI]

# Review of algebraic fractions



If f($x$) and g($x$) are polynomials, the expression $\frac{f(x)}{g(x)}$ is an *algebraic fraction* or *rational function.* It may also be called a *rational expression.* There are many occasions in mathematics when a problem reduces to the manipulation of algebraic fractions, and the rules for this are exactly the same as those for numerical fractions.

## Simplifying fractions

To simplify a fraction, you look for a factor common to both the numerator (top line) and the denominator (bottom line) and cancel by it.

For example, in arithmetic

$$\frac{15}{20} = \frac{5 \times 3}{5 \times 4} = \frac{3}{4}$$

and in algebra

$$\frac{6a}{9a^2} = \frac{2 \times 3 \times a}{3 \times 3 \times a \times a} = \frac{2}{3a}.$$

Notice how you must *factorise* both the numerator and denominator before cancelling, since it is only possible to cancel by a *common factor.* In some cases this involves putting brackets in:

$$\frac{2a+4}{a^2-4} = \frac{2(a+2)}{(a+2)(a-2)} = \frac{2}{(a-2)}.$$

## Multiplying and dividing fractions

Multiplying fractions involves cancelling any factors common to the numerator and denominator. For example:

$$\frac{10a}{3b^2} \times \frac{9ab}{25} = \frac{2 \times 5 \times a}{3 \times b \times b} \times \frac{3 \times 3 \times a \times b}{5 \times 5} = \frac{6a^2}{5b}.$$

As with simplifying, it is often necessary to factorise any algebraic expressions first:

$$\frac{a^2+3a+2}{9} = \frac{12}{a+1} = \frac{(a+1)(a+2)}{3 \times 3} \times \frac{3 \times 4}{(a+1)}$$

$$= \frac{(a+2)}{3} \times \frac{4}{1}$$

$$= \frac{4(a+2)}{3}.$$

Remember that when one fraction is divided by another, you change $\div$ to $\times$ and invert the fraction which follows the $\div$ symbol. For example:

$$\frac{12}{x^2-1} \div \frac{4}{x+1} = \frac{12}{(x+1)(x-1)} \times \frac{(x+1)}{4}$$

$$= \frac{3}{(x-1)}.$$

## Addition and subtraction of fractions

To add or subtract two fractions they must be replaced by equivalent fractions, both of which have the same denominator.

For example:

$$\frac{2}{3} + \frac{1}{4} = \frac{8}{12} + \frac{3}{12} = \frac{11}{12}.$$

Similarly, in algebra:

$$\frac{2x}{3} + \frac{x}{4} = \frac{8x}{12} + \frac{3x}{12} = \frac{11x}{12}$$

and $$\frac{2}{3x} + \frac{1}{4x} = \frac{8}{12x} + \frac{3}{12x} = \frac{11}{12x}.$$

Notice how you only need $12x$ here, not $12x^2$

You must take particular care when the subtraction of fractions introduces a sign change. For example:

$$\frac{4x-3}{6} - \frac{2x+1}{4} = \frac{2(4x-3)-3(2x+1)}{12}$$

$$= \frac{8x-6-6x-3}{12}$$

$$= \frac{2x-9}{12}.$$

Notice how in addition and subtraction, the new denominator is the *lowest common multiple* of the original denominators. When two denominators have no *common factor*, their product gives the new denominator. For example:

$$\frac{2}{y+3} + \frac{3}{y-2} = \frac{2(y-2)+3(y+3)}{(y+3)(y-2)}$$

$$= \frac{2y-4+3y+9}{(y+3)(y-2)}$$

$$= \frac{5y+5}{(y+3)(y-2)}$$

$$= \frac{5(y+1)}{(y+3)(y-2)}.$$

It may be necessary to factorise denominators in order to identify common factors, as shown here:

$$\frac{2b}{a^2-b^2}-\frac{3}{a+b}=\frac{2b}{(a+b)(a-b)}-\frac{3}{(a+b)}$$

$$=\frac{2b-3(a-b)}{(a+b)(a-b)}$$

$(a+b)$ is a common factor

$$=\frac{5b-3a}{(a+b)(a-b)}.$$

## EXERCISE 1B

*Simplify the expressions in questions 1–10.*

**1** $\frac{6a}{b}\times\frac{a}{9b^2}$

**2** $\frac{5xy}{3}\div 15xy^2$

**3** $\frac{x^2-9}{x^2-9x+18}$

**4** $\frac{5x-1}{x+3}\times\frac{x^2+6x+9}{5x^2+4x-1}$

**5** $\frac{4x^2-25}{4x^2+20x+25}$

**6** $\frac{a^2+a-12}{5}\times\frac{3}{4a-12}$

**7** $\frac{4x^2-9}{x^2+2x+1}\div\frac{2x-3}{x^2+x}$

**8** $\frac{2p+4}{5}\div(p^2-4)$

**9** $\frac{a^2-b^2}{2a^2+ab-b^2}$

**10** $\frac{x^2+8x+16}{x^2+6x+9}\times\frac{x^2+2x-3}{x^2+4x}$

*In questions 11–24 write each of the expressions as a single fraction in its simplest form.*

**11** $\frac{1}{4x}+\frac{1}{5x}$

**12** $\frac{x}{3}-\frac{(x+1)}{4}$

**13** $\frac{a}{a+1}+\frac{1}{a-1}$

**14** $\frac{2}{x-3}+\frac{3}{x-2}$

**15** $\frac{x}{x^2-4}-\frac{1}{x+2}$

**16** $\frac{p^2}{p^2-1}-\frac{p^2}{p^2+1}$

**17** $\frac{2}{a+1}-\frac{a}{a^2+1}$

**18** $\frac{2y}{(y+2)^2}-\frac{4}{y+4}$

**19** $x+\frac{1}{x+1}$

**20** $\frac{2}{b^2+2b+1}-\frac{3}{b+1}$

**21** $\frac{2}{3(x-1)}+\frac{3}{2(x+1)}$

**22** $\frac{6}{5(x+2)}-\frac{2x}{(x+2)^2}$

**23** $\frac{2}{a+2}-\frac{a-2}{2a^2+a-6}$

**24** $\frac{1}{x-2}+\frac{1}{x}+\frac{1}{x+2}$

## Equations involving algebraic fractions

The easiest way to solve an equation involving fractions is usually to multiply both sides by an expression which will cancel out the fractions.

**EXAMPLE 1.6**

Solve $\frac{x}{3} + \frac{2x}{5} = 4$.

**SOLUTION**

Multiplying by 15 (the lowest common multiple of 3 and 5) gives

$$15 \times \frac{x}{3} + 15 \times \frac{2x}{5} = 15 \times 4$$

Notice that all three terms must be multiplied by 15

$$\Rightarrow \quad 5x + 6x = 60$$

$$\Rightarrow \quad 11x = 60$$

$$\Rightarrow \quad x = \frac{60}{11}.$$

A similar method applies when the denominator is algebraic.

**EXAMPLE 1.7**

Solve $\frac{5}{x} - \frac{4}{x+1} = 1$.

**SOLUTION**

Multiplying by $x(x + 1)$ (the least common multiple of $x$ and $x + 1$) gives

$$\frac{5x(x+1)}{x} - \frac{4x(x+1)}{x+1} = x(x+1)$$

$$\Rightarrow \quad 5(x+1) - 4x = x(x+1)$$

$$\Rightarrow \quad 5x + 5 - 4x = x^2 + x$$

$$\Rightarrow \quad x^2 = 5$$

$$\Rightarrow \quad x = \pm\sqrt{5}.$$

In Example 1.7, the lowest common multiple of the denominators is their product, but this is not always the case.

**EXAMPLE 1.8**

Solve $\dfrac{1}{(x-3)(x-1)}+\dfrac{1}{x(x-1)}=-\dfrac{1}{x(x-3)}$.

**SOLUTION**

Here you only need to multiply by $x(x-3)(x-1)$ to eliminate all the fractions. This gives

$$\frac{x(x-3)(x-1)}{(x-3)(x-1)}+\frac{x(x-3)(x-1)}{x(x-1)}=\frac{-x(x-3)(x-1)}{x(x-3)}$$

$$\Rightarrow \quad x+(x-3)=-(x-1)$$

$$\Rightarrow \quad 2x-3=-x+1$$

$$\Rightarrow \quad 3x=4$$

$$\Rightarrow \quad x=\tfrac{4}{3}.$$

Fractional algebraic equations arise in a number of situations, including, as in the following example, problems connecting distance, speed and time. The relationship

$$\text{time}=\frac{\text{distance}}{\text{speed}}$$

is useful here.

**EXAMPLE 1.9**

Each day I travel 10 km from home to work. One day, because of road works, my average speed was 5 km $h^{-1}$ slower than usual, and my journey took an extra 10 minutes.

Taking $x$ km $h^{-1}$ as my usual speed:

**(i)** write down an expression in $x$ which represents my usual time in hours;

**(ii)** write down an expression in $x$ which represents my time when I travelled 5 km $h^{-1}$ slower than usual;

**(iii)** use these expressions to form an equation in $x$ and solve it.

**(iv)** How long does my journey usually take?

**SOLUTION**

**(i)** $\text{Time}=\dfrac{\text{distance}}{\text{speed}} \quad\Rightarrow\quad \text{usual time}=\dfrac{10}{x}$.

**(ii)** I now travel at $(x-5)$ km $h^{-1}$, so the longer time $=\dfrac{10}{x-5}$.

**(iii)** The difference in these times is 10 minutes, or $\frac{1}{6}$ hour, so

$$\frac{10}{x-5}-\frac{10}{x}=\frac{1}{6}.$$

Multiplying by $6x(x-5)$ gives

$$\frac{60x(x-5)}{(x-5)} - \frac{60x(x-5)}{x} = \frac{6x(x-5)}{6}$$

$$\Rightarrow \quad 60x - 60(x-5) = x(x-5)$$

$$\Rightarrow \quad 60x - 60x + 300 = x^2 - 5x$$

$$\Rightarrow \quad x^2 - 5x - 300 = 0$$

$$\Rightarrow \quad (x-20)(x+15) = 0$$

$$\Rightarrow \quad x = 20 \text{ or } x = -15.$$

**(iv)** Reject $x = -15$, since $x$ km h$^{-1}$ is a speed.
Usual speed = 20 km h$^{-1}$.
Usual time = $\frac{10}{x}$ hours = 30 minutes.

## EXERCISE 1C

**1** Solve the following equations.

**(i)** $\frac{2x}{7} - \frac{x}{4} = 3$

**(ii)** $\frac{5}{4x} + \frac{3}{2x} = \frac{11}{16}$

**(iii)** $\frac{2}{x} - \frac{5}{2x-1} = 0$

**(iv)** $x - 3 = \frac{x+2}{x-2}$

**(v)** $\frac{1}{x} + x + 1 = \frac{13}{3}$

**(vi)** $\frac{2x}{x+1} - \frac{1}{x-1} = 1$

**(vii)** $\frac{x}{x-1} - \frac{x-1}{x} = 2$

**2** I have £6 to spend on crisps for a party. When I get to the shop I find that the price has been reduced by 1 penny per packet, and I can buy one packet more than I expected. Taking $x$ pence as the original cost of a packet of crisps:

**(i)** write down an expression in $x$ which represents the number of packets that I expected to buy;

**(ii)** write down an expression in $x$ which represents the number of packets bought at the reduced price;

**(iii)** form an equation in $x$ and solve it to find the original cost.

**3** The distance from Manchester to Oxford is 270 km. One day, road works on the M6 meant that my average speed was 10 km h$^{-1}$ less than I had anticipated, and so I arrived 18 minutes later than planned. Taking $x$ km h$^{-1}$ as the anticipated average speed:

**(i)** write down an expression in $x$ for the anticipated and actual times of the journey;

**(ii)** form an equation in $x$ and solve it;

**(iii)** find the time of my arrival in Oxford if I left home at 10 am.

**4** Each time someone leaves the firm of Honeys, he or she is taken out for a meal by the rest of the staff. On one such occasion the bill came to £272, and each member of staff remaining with the firm paid £1 extra to cover the cost of the meal for the one who was leaving. Taking £$x$ as the cost of the meal, write down an equation in $x$ and solve it.

How many staff were left working for Honeys?

**5** A Swiss roll cake is 21 cm long. When I cut it into slices, I can get two extra slices if I reduce the thickness of each slice by $\frac{1}{4}$ cm. Taking $x$ as the number of thicker slices, write down an equation in $x$ and solve it.

**6** Two electrical resistances may be connected in series or in parallel. In series, the equivalent single resistance is the sum of the two resistances, but in parallel, the two resistances $R_1$ and $R_2$ are equivalent to a single resistance $R$ where

$$\frac{1}{R_1} + \frac{1}{R_2} = \frac{1}{R}.$$

**(i)** Find the single resistance which is equivalent to resistances of 3 and 4 ohms connected in parallel.

**(ii)** What resistance must be added in parallel to a resistance of 6 ohms to give a resistance of 2.4 ohms?

**(iii)** What is the effect of connecting two equal resistances in parallel?

**INVESTIGATION**

Investigate

$$\frac{x}{x-1} + \frac{x-1}{x} = 2$$

and give an explanation for what happens.

## Partial fractions

Until this point, any instruction to simplify an algebraic fractional expression was asking you to give the expansion as a single fraction. Sometimes, however, it is easier to deal with two or three simple separate fractions than it is to handle one more complicated one. This is the case when you are using the binomial theorem to obtain a series expansion.

For example:

$$\frac{1}{(1+2x)(1+x)}$$

may be written as

$$\frac{2}{(1+2x)} - \frac{1}{(1+x)}.$$

The two-part expression $\frac{2}{(1+2x)} - \frac{1}{(1+x)}$ is much easier to expand than $\frac{1}{(1+2x)(1+x)}$.

When integrating, it is even more important to work with a number of simple fractions than to combine them into one. For example, the only analytical method for integrating $\frac{1}{(1+2x)(1+x)}$ first involves writing it as $\frac{2}{(1+2x)} - \frac{1}{(1+x)}$. You will meet this application in Chapter 3.

This process of taking an expression such as $\frac{1}{(1+2x)(1+x)}$ and writing it in the form $\frac{2}{(1+2x)} - \frac{1}{(1+x)}$ is called expressing the algebraic fraction in *partial fractions.*

How can this be done in general?

It is sufficient to consider only *proper* algebraic fractions, that is fractions where the order of the numerator (top line) is strictly less than that of denominator (bottom line). The following, for example, are proper fractions:

$$\frac{2}{1+x}, \frac{5x-1}{x^2-3}, \frac{7x}{(x+1)(x-2)}.$$

Examples of improper fractions are

$$\frac{2x}{x+1} \qquad \left(\text{which can be written as } 2 - \frac{2}{x+1}\right)$$

and

$$\frac{x^2}{x-2} \qquad \left(\text{which can be written as } x + 2 + \frac{4}{x-2}\right).$$

It can be shown that, when a proper algebraic fraction is decomposed into its partial fractions, each of the partial fractions will be a proper fraction.

When finding partial fractions you must always assume the most general numerator possible, and the method for doing this is illustrated in the following examples.

## Type 1: denominators of the form $(ax + b)(cx + d)$



**EXAMPLE 1.10**

Express $\dfrac{4 + x}{(1 + x)(2 - x)}$ as a sum of partial fractions.

Remember: a linear denominator $\Rightarrow$ a constant numerator if the fraction is to be a proper fraction

**SOLUTION**

Assume

$$\frac{4 + x}{(1 + x)(2 - x)} \equiv \frac{A}{1 + x} + \frac{B}{2 - x}$$

Multiplying both sides by $(1 + x)(2 - x)$ gives

$$4 + x \equiv A(2 - x) + B(1 + x). \qquad ①$$

This is an identity; it is true for all values of $x$.

There are two possible ways in which you can find the constants $A$ and $B$. You can either

- substitute *any two* values of $x$ in ① (two values are needed to give two equations to solve for the two unknowns $A$ and $B$); or
- equate the constant terms to give one equation (this is the same as putting $x = 0$) and the coefficients of $x$ to give another.

Sometimes one method is easier than the other, and in practice you will often want to use a combination of the two.

### *Method 1: Substitution*

Although you can substitute any two values of $x$, the easiest to use are $x = 2$ and $x = -1$, since each makes the value of one bracket 0 in the identity

$$4 + x \equiv A(2 - x) + B(1 + x)$$

$$x = 2 \quad \Rightarrow \quad 4 + 2 = A(2 - 2) + B(1 + 2)$$

$$6 = 3B \quad \Rightarrow \quad B = 2$$

$$x = -1 \quad \Rightarrow \quad 4 - 1 = A(2 + 1) + B(1 - 1)$$

$$3 = 3A \quad \Rightarrow \quad A = 1.$$

Substituting these values for $A$ and $B$ gives

$$\frac{4 + x}{(1 + x)(2 - x)} \equiv \frac{1}{1 + x} + \frac{2}{2 - x}.$$

1

Algebra

*Method 2: Equating coefficients*

In this method, you write the right-hand side of

$$4 + x \equiv A(2 - x) + B(1 + x)$$

as a polynomial in $x$, and then compare the coefficients of the various terms:

$$4 + x \equiv 2A - Ax + B + Bx$$

$$4 + 1x \equiv (2A + B) + (-A + B)x.$$

Equating the constant terms: $4 = 2A + B$.

These are simultaneous equations in $A$ and $B$

Equating the coefficients of $x$: $1 = -A + B$.

Solving these simultaneous equations gives $A = 1$ and $B = 2$ as before.

---

? In each of these methods the identity ($\equiv$) was later replaced by equality ($=$). Why did this happen?

---

In some cases it is necessary to factorise the denominator before finding the partial fractions.

**EXAMPLE 1.11**

Express $\dfrac{2}{4 - x^2}$ as a sum of partial fractions.

**SOLUTION**

$$\frac{2}{4 - x^2} \equiv \frac{2}{(2 + x)(2 - x)}$$

$$\equiv \frac{A}{2 + x} + \frac{B}{2 - x}.$$

Multiplying both sides by $(2 + x)(2 - x)$ gives

$$2 \equiv A(2 - x) + B(2 + x) \qquad ①$$

$$2 \equiv (2A + 2B) + x(B - A).$$

Equating constant terms: $2 = 2A + 2B$. ②

Equating coefficients of $x$: $0 = B - A$, so $B = A$.

Substituting in ② gives $A = B = \frac{1}{2}$.

Using these values

$$\frac{2}{(2 + x)(2 - x)} \equiv \frac{\frac{1}{2}}{2 + x} + \frac{\frac{1}{2}}{2 - x}$$

$$\equiv \frac{1}{2(2 + x)} + \frac{1}{2(2 - x)}.$$

**EXERCISE 1D**

*Express each of the following fractions as a sum of partial fractions.*

**1** $\dfrac{5}{(x-2)(x+3)}$

**2** $\dfrac{1}{x(x+1)}$

**3** $\dfrac{6}{(x-1)(x-4)}$

**4** $\dfrac{x+5}{(x-1)(x+2)}$

**5** $\dfrac{3x}{(2x-1)(x+1)}$

**6** $\dfrac{4}{x^2-2x}$

**7** $\dfrac{2}{(x-1)(3x-1)}$

**8** $\dfrac{x-1}{x^2-3x-4}$

**9** $\dfrac{x+2}{2x^2-x}$

**10** $\dfrac{7}{2x^2+x-6}$

**11** $\dfrac{2x-1}{2x^2+3x-20}$

**12** $\dfrac{2x+5}{18x^2-8}$

## Type 2: denominators of the form $(ax + b)(cx^2 + d)$

**EXAMPLE 1.12**

Express $\dfrac{2x+3}{(x-1)(x^2+4)}$ as a sum of partial fractions.

**SOLUTION**

You need to assume a numerator of order 1 for the partial fraction with a denominator of $x^2 + 4$, which is of order 2.

$$\frac{2x+3}{(x-1)(x^2+4)} \equiv \frac{A}{x-1} + \frac{Bx+C}{x^2+4}$$

$Bx + C$ is the most general numerator of order 1

Multiplying both sides by $(x-1)(x^2+4)$ gives

$$2x + 3 \equiv A(x^2+4) + (Bx+C)(x-1) \qquad ①$$

$$x = 1 \quad \Rightarrow \quad 5 = 5A \quad \Rightarrow \quad A = 1.$$

The other two unknowns, $B$ and $C$, are most easily found by equating coefficients. Identity ① may be rewritten as

$$2x + 3 \equiv (A+B)x^2 + (-B+C)x + (4A-C).$$

Equating coefficients of $x^2$: $\quad 0 = A + B \quad \Rightarrow \quad B = -1.$

Equating constant terms: $\quad 3 = 4A - C \quad \Rightarrow \quad C = 1.$

This gives

$$\frac{2x+3}{(x-1)(x^2+4)} \equiv \frac{1}{x-1} + \frac{1-x}{x^2+4}.$$

*Note*

Notice how Example 1.12 uses a combination of the two methods.

### Type 3: denominators of the form $(ax + b)(cx + d)^2$

The factor $(cx + d)^2$ is of order 2, so it would have an order 1 numerator in the partial fractions. However, in the case of a repeated factor there is a simpler form.

Consider $\dfrac{4x + 5}{(2x + 1)^2}$

This can be written as

$$\frac{2(2x + 1) + 3}{(2x + 1)^2}$$

$$\equiv \frac{2(2x + 1)}{(2x + 1)^2} + \frac{3}{(2x + 1)^2}$$

$$\equiv \frac{2}{(2x + 1)} + \frac{3}{(2x + 1)^2}.$$

***Note***

In this form, both the numerators are constant.

In a similar way, any fraction of the form $\dfrac{px + q}{(cx + d)^2}$ can be written as

$$\frac{A}{(cx + d)} + \frac{B}{(cx + d)^2}.$$

When expressing an algebraic fraction in partial fractions, you are aiming to find the simplest partial fractions possible, so you would want the form where the numerators are constant.

**EXAMPLE 1.13**

Express $\dfrac{x + 1}{(x - 1)(x - 2)^2}$ as a sum of partial fractions.

**SOLUTION**

Let $\dfrac{x + 1}{(x - 1)(x - 2)^2} \equiv \dfrac{A}{(x - 1)} + \dfrac{B}{(x - 2)} + \dfrac{C}{(x - 2)^2}$.

Notice that you only need $(x - 2)^2$ here and not $(x - 2)^3$

Multiplying both sides by $(x - 1)(x - 2)^2$ gives

$$x + 1 \equiv A(x - 2)^2 + B(x - 1)(x - 2) + C(x - 1).$$

$x = 1$ (so that $x - 1 = 0$) $\Rightarrow 2 = A(-1)^2 \Rightarrow A = 2$

$x = 2$ (so that $x - 2 = 0$) $\Rightarrow 3 = C$.

Equating coefficients of $x^2$ $\Rightarrow 0 = A + B \Rightarrow B = -2$.

This gives

$$\frac{x + 1}{(x - 1)(x - 2)^2} \equiv \frac{2}{x - 1} - \frac{2}{x - 2} + \frac{3}{(x - 2)^2}.$$

**EXAMPLE 1.14**

Express $\dfrac{5x^2-3}{x^2(x+1)}$ as a sum of partial fractions.

**SOLUTION**

Let $\quad \dfrac{5x^2-3}{x^2(x+1)} \equiv \dfrac{A}{x} + \dfrac{B}{x^2} + \dfrac{C}{x+1}$.

Multiplying both sides by $x^2(x+1)$ gives

$$5x^2 - 3 \equiv Ax(x+1) + B(x+1) + Cx^2.$$

$$x = 0 \quad \Rightarrow \quad -3 = B$$

$$x = -1 \quad \Rightarrow \quad +2 = C.$$

Equating coefficients of $x^2$: $\quad +5 = A + C \quad \Rightarrow \quad A = 3.$

This gives

$$\frac{5x^2-3}{x^2(x+1)} \equiv \frac{3}{x} - \frac{3}{x^2} + \frac{2}{x+1}.$$

**EXERCISE 1E**

*Express each of the following fractions as a sum of partial fractions.*

**1** $\dfrac{4}{(1-3x)(1-x)^2}$ **2** $\dfrac{4+2x}{(2x-1)(x^2+1)}$ **3** $\dfrac{5-2x}{(x-1)^2(x+2)}$

**4** $\dfrac{2x+1}{(x-2)(x^2+4)}$ **5** $\dfrac{2x^2+x+4}{(2x^2-3)(x+2)}$ **6** $\dfrac{x^2-1}{x^2(2x+1)}$

**7** $\dfrac{x^2+3}{x(3x^2-1)}$ **8** $\dfrac{2x^2+x+2}{(2x^2+1)(x+1)}$ **9** $\dfrac{4x^2-3}{x(2x-1)^2}$

**10** Given that

$$\frac{x^2+2x+7}{(2x+3)(x^2+4)} \equiv \frac{A}{(2x+3)} + \frac{Bx+C}{(x^2+4)}$$

find the values of the constants $A$, $B$ and $C$.

**[MEI, part]**

**11** Calculate the values of the constants $A$, $B$ and $C$ for which

$$\frac{x^2-4x+23}{(x-5)(x^2+3)} \equiv \frac{A}{(x-5)} + \frac{Bx+C}{(x^2+3)}.$$

**[MEI, part]**

## Using partial fractions with the binomial expansion

One of the most common reasons for writing an expression in partial fractions is to enable binomial expansions to be applied, as in the following example.

**EXAMPLE 1.15**

Express $\frac{2x+7}{(x-1)(x+2)}$ in partial fractions and hence find the first three terms of its binomial expansion, stating the values of $x$ for which this is valid.

**SOLUTION**

$$\frac{2x+7}{(x-1)(x+2)} \equiv \frac{A}{(x-1)} + \frac{B}{(x+2)}$$

Multiplying both sides by $(x-1)(x+2)$ gives

$$2x+7 \equiv A(x+2) + B(x-1).$$

$$x = 1 \quad \Rightarrow \quad 9 = 3A \quad \Rightarrow \quad A = 3$$

$$x = -2 \quad \Rightarrow \quad 3 = -3B \quad \Rightarrow \quad B = -1.$$

This gives

$$\frac{2x+7}{(x-1)(x+2)} \equiv \frac{3}{(x-1)} - \frac{1}{(x+2)}.$$

In order to obtain the binomial expansion, each bracket must be of the form $(1 \pm \ldots)$, giving

$$\frac{2x+7}{(x-1)(x+2)} \equiv \frac{-3}{(1-x)} - \frac{1}{2\left(1+\frac{x}{2}\right)}$$

$$\equiv -3(1-x)^{-1} - \frac{1}{2}\left(1+\frac{x}{2}\right)^{-1}. \quad ①$$

The two binomial expansions are

$$(1-x)^{-1} = 1 + (-1)(-x) + \frac{(-1)(-2)}{2!}(-x)^2 + \ldots \quad \text{for } |x| < 1$$

$$\approx 1 + x + x^2$$

and $$\left(1+\frac{x}{2}\right)^{-1} = 1 + (-1)\left(\frac{x}{2}\right) + \frac{(-1)(-2)}{2!}\left(\frac{x}{2}\right)^2 + \ldots \quad \text{for } \left|\frac{x}{2}\right| < 1$$

$$\approx 1 - \frac{x}{2} + \frac{x^2}{4}.$$

Substituting these in ① gives

$$\frac{2x+7}{(x-1)(x+2)} \approx -3(1+x+x^2) - \frac{1}{2}\left(1 - \frac{x}{2} + \frac{x^2}{4}\right)$$

$$= -\frac{7}{2} - \frac{11x}{4} - \frac{25x^2}{8}.$$

The expansion is valid when $|x| < 1$ and $\left|\frac{x}{2}\right| < 1$. The stricter of these is $|x| < 1$

## INVESTIGATION

Find a binomial expansion for the function

$$f(x) = \frac{1}{(1+2x)(1-x)}$$

and state the values of $x$ for which it is valid:

**(i)** by writing it as $(1+2x)^{-1}(1-x)^{-1}$;
**(ii)** by writing it as $[1+(x-2x^2)]^{-1}$ and treating $(x-2x^2)$ as one term;
**(iii)** by first expressing $f(x)$ as a sum of partial fractions.

Decide which method you find simplest for the following cases:
**(a)** when a linear approximation for $f(x)$ is required;
**(b)** when a quadratic approximation for $f(x)$ is required;
**(c)** when the coefficient of $x^n$ is required.

## EXERCISE 1F

*Find the first three terms in the binomial expansion of the fractions in questions* 1 – 4.

**1** $\dfrac{4}{(1-3x)(1-x)^2}$

**2** $\dfrac{4+2x}{(2x-1)(x^2+1)}$

**3** $\dfrac{5-2x}{(x-1)^2(x+2)}$

**4** $\dfrac{2x+1}{(x-2)(x^2+4)}$

**5 (i)** Express $\dfrac{7-4x}{(2x-1)(x+2)}$ in partial fractions as $\dfrac{A}{(2x-1)} + \dfrac{B}{(x+2)}$ where $A$ and $B$ are to be found.

**(ii)** Find the expansion of $\dfrac{1}{(1-2x)}$ in the form $a + bx + cx^2 + \ldots$ where $a$, $b$ and $c$ are to be found. Give the range of values of $x$ for which this expansion is valid.

**(iii)** Find the expansion of $\dfrac{1}{(2+x)}$ as far as the term containing $x^2$. Give the range of values of $x$ for which this expansion is valid.

**(iv)** Hence find a quadratic approximation for $\dfrac{7-4x}{(2x-1)(x+2)}$ when $|x|$ is small.

Find the relative error in this approximation when $x = 0.1$.

[MEI]

## KEY POINTS

**1** The general binomial expansion for $n \in \mathbb{R}$ is

$$(1+x)^n = 1 + nx + \frac{n(n-1)}{2!}x^2 + \frac{n(n-1)(n-2)}{3!}x^3 + \dots$$

In the special case when $n \in \mathbb{N}$, the series expansion is finite and valid for all $x$.

When $n \notin \mathbb{N}$, the series expansion is non-terminating (infinite) and valid only if $|x| < 1$.

**2** When $n \notin \mathbb{N}$, $(a+x)^n$ should be written as $a^n\left(1+\frac{x}{a}\right)^n$ before obtaining the binomial expansion.

**3** When multiplying algebraic fractions, you can only cancel when the same factor occurs in both the numerator and denominator.

**4** When adding or subtracting algebraic fractions, you first need to find a common denominator.

**5** The easiest way to solve any equation involving fractions is usually to multiply both sides by a quantity which will eliminate the fractions.

**6** A proper algebraic fraction with a denominator which factorises can be decomposed into a sum of proper partial fractions.

**7** The following forms of partial fraction should be used:

$$\frac{px+q}{(ax+b)(cx+d)} \equiv \frac{A}{ax+b} + \frac{B}{cx+d}$$

$$\frac{px^2+qr+r}{(ax+b)(cx^2+d)} \equiv \frac{A}{ax+b} + \frac{Bx+C}{cx^2+d}$$

$$\frac{px^2+qx+r}{(ax+b)(cx+d)^2} \equiv \frac{A}{ax+b} + \frac{B}{cx+d} + \frac{C}{(cx+d)^2}.$$

# 2 Trigonometry



**Music, when soft voices die,**
**Vibrates in the memory –**

*P.B. Shelley*

Both of these photographs show forms of waves. In each case, estimate the wavelength and the amplitude in metres (see figure 2.1).

Use your measurements to suggest, for each curve, values of $a$ and $b$ which would make $y = a\sin bx$ a suitable model for the curve.

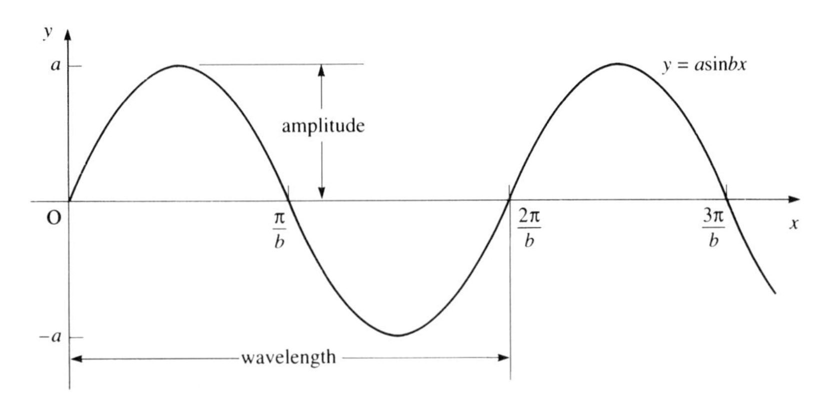


**Figure 2.1**

## Introduction to further trigonometry

The photographs at the start of this chapter show just two of the countless examples of waves and oscillations that are part of the world around us.

Because such phenomena are modelled by trigonometric (and especially sine and cosine) functions, trigonometry has an importance in mathematics far beyond its origins in right-angled triangles.

In this chapter, we develop a number of techniques for manipulating trigonometric functions and equations, starting with the compound-angle formulae.

## Compound-angle formulae

ACTIVITY

Find an acute angle $\theta$ so that $\sin(\theta + 60°) = \cos(\theta - 60°)$.

**Hint:** Try drawing graphs and searching for a numerical solution.

You should be able to find the solution using either of these methods, but replacing 60° by, for example, 35° would make both of these methods rather tedious. In this chapter you will meet some formulae which help you to solve such equations more efficiently.

It is tempting to think that $\sin(\theta + 60°)$ should equal $\sin\theta + \sin 60°$, but this is not so, as you can see by substituting a numerical value of $\theta$. For example, putting $\theta$ = 30° gives $\sin(\theta + 60°) = 1$, but $\sin\theta + \sin 60° \approx 1.366$.

To find an expression for $\sin(\theta + 60°)$, you would use the *compound-angle formula*:

$$\sin(\theta + \phi) = \sin\theta \cos\phi + \cos\theta \sin\phi.$$

This is proved below in the case when $\theta$ and $\phi$ are acute angles. It is, however, true for all values of the angles. It is an *identity*.

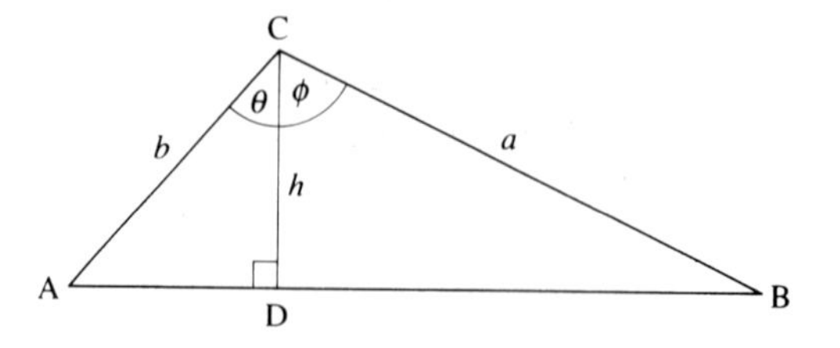


**Figure 2.2**

Using the trigonometric formula for the area of a triangle in figure 2.2:

$$\text{area ABC} = \text{area ADC} + \text{area DBC}$$

$$\tfrac{1}{2}ab\sin(\theta + \phi) = \tfrac{1}{2}bh\sin\theta + \tfrac{1}{2}ah\sin\phi$$

$h = a\cos\phi$ from $\triangle$DBC

$h = b\cos\theta$ from $\triangle$ADC

$$ab\sin(\theta + \phi) = ab\sin\theta\cos\phi + ab\cos\theta\sin\phi$$

which gives

$$\sin(\theta + \phi) = \sin\theta\cos\phi + \cos\theta\sin\phi \qquad ①$$

This is the first of the compound-angle formulae (or expansions), and it can be used to prove several more. These are true for all values of $\theta$ and $\phi$.

Replacing $\phi$ by $-\phi$ in ① gives

$$\sin(\theta - \phi) = \sin\theta\cos(-\phi) + \cos\theta\sin(-\phi)$$

$\cos(-\phi) = \cos\phi$

$\sin(-\phi) = -\sin\phi$

$$\Rightarrow \quad \sin(\theta - \phi) = \sin\theta\cos\phi - \cos\theta\sin\phi. \qquad ②$$

**ACTIVITY**

Derive the rest of these formulae.

**(i)** To find an expansion for $\cos(\theta - \phi)$ replace $\theta$ by $(90° - \theta)$ in the expansion of $\sin(\theta + \phi)$.

**Hint:** $\sin(90° - \theta) = \cos\theta$ and $\cos(90° - \theta) = \sin\theta$

**(ii)** To find an expansion for $\cos(\theta + \phi)$ replace $\phi$ by $(-\phi)$ in the expansion of $\cos(\theta - \phi)$.

**(iii)** To find an expansion for $\tan(\theta + \phi)$, write $\tan(\theta + \phi) = \dfrac{\sin(\theta + \phi)}{\cos(\theta + \phi)}$.

**Hint:** After using the expansions of $\sin(\theta + \phi)$ and $\cos(\theta + \phi)$, divide the numerator and denominator of the resulting fraction by $\cos\theta\cos\phi$ to give an expansion in terms of $\tan\theta$ and $\tan\phi$.

**(iv)** To find an expansion for $\tan(\theta - \phi)$ in terms of $\tan\theta$ and $\tan\phi$, replace $\phi$ by $(-\phi)$ in the expansion of $\tan(\theta + \phi)$.

The four results obtained in the activity, together with the two previous results, form the set of compound-angle formulae:

$$\sin(\theta + \phi) = \sin\theta\cos\phi + \cos\theta\sin\phi$$

$$\sin(\theta - \phi) = \sin\theta\cos\phi - \cos\theta\sin\phi$$

$$\cos(\theta + \phi) = \cos\theta\cos\phi - \sin\theta\sin\phi$$

$$\cos(\theta - \phi) = \cos\theta\cos\phi + \sin\theta\sin\phi$$

$$\tan(\theta + \phi) = \frac{\tan\theta + \tan\phi}{1 - \tan\theta\tan\phi}$$

$$\tan(\theta - \phi) = \frac{\tan\theta - \tan\phi}{1 + \tan\theta\tan\phi}.$$

You are now in a position to solve the earlier problem more easily. To find an acute angle $\theta$ such that $\sin(\theta + 60°) = \cos(\theta - 60°)$, you expand each side using the compound-angle formulae:

$$\sin(\theta + 60°) = \sin\theta\cos60° + \cos\theta\sin60°$$

$$= \frac{1}{2}\sin\theta + \frac{\sqrt{3}}{2}\cos\theta \qquad ①$$

$$\cos(\theta - 60°) = \cos\theta\cos60° + \sin\theta\sin60°$$

$$= \frac{1}{2}\cos\theta + \frac{\sqrt{3}}{2}\sin\theta. \qquad ②$$

From ① and ②

$$\frac{1}{2}\sin\theta + \frac{\sqrt{3}}{2}\cos\theta = \frac{1}{2}\cos\theta + \frac{\sqrt{3}}{2}\sin\theta$$

$$\sin\theta + \sqrt{3}\cos\theta = \cos\theta + \sqrt{3}\sin\theta.$$

Collect like terms:

$$\Rightarrow \quad (\sqrt{3} - 1)\cos\theta = (\sqrt{3} - 1)\sin\theta$$

$$\cos\theta = \sin\theta.$$

Divide by $\cos\theta$:

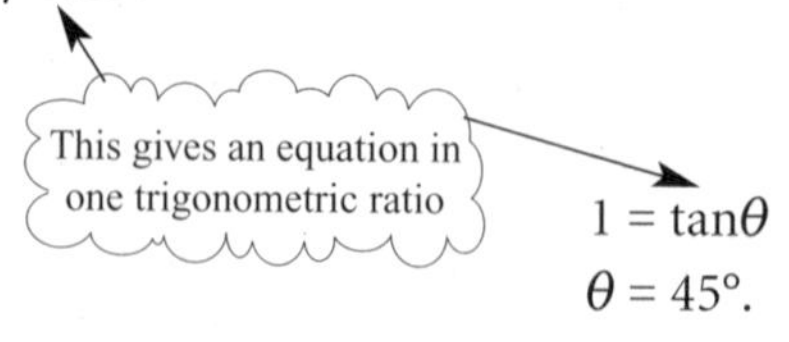


$$1 = \tan\theta$$

$$\theta = 45°.$$

Since an acute angle was required, this is the only root.

### When do you use the compound-angle formulae?

You have already seen compound-angle formulae used in solving a trigonometric equation and this is quite a common application of them. However, their significance goes well beyond that since they form the basis for a number of important techniques. Those covered in this book are:

- **Derivation of double-angle formulae**
  The derivation and uses of these are covered on pages 33 to 36.
- **Addition of different sine and cosine functions**
  This is covered on pages 37 to 40 and 41 to 44 of this chapter. The work on pages 37 to 40 is non-syllabus material. Because the basic wave form is a sine curve, it has many applications in applied mathematics, physics, chemistry, etc.
- **Differentiation of trigonometric functions**
  This is covered in Chapter 3 and depends on using either the compound-angle formulae or the factor formulae which are derived from them.

You will see from this that the compound-angle formulae are important in the development of the subject. Some people learn them by heart, others think it is safer to look them up when they are needed. Whichever policy you adopt, you should understand these formulae and recognise their form. Without that you will be unable to do the next example, which uses one of them in reverse.

**EXAMPLE 2.1**

Simplify $\cos\theta\cos3\theta - \sin\theta\sin3\theta$.

**SOLUTION**

The formula which has the same pattern of cos cos – sin sin is

$$\cos(\theta + \phi) = \cos\theta\cos\phi - \sin\theta\sin\phi$$

Using this, and replacing $\phi$ by $3\theta$, gives

$$\begin{aligned}\cos\theta\cos3\theta - \sin\theta\sin3\theta &= \cos(\theta + 3\theta)\\ &= \cos4\theta.\end{aligned}$$

**EXERCISE 2A**

**1** Use the compound-angle formulae to write the following as surds.

**(i)** $\sin75° = \sin(45° + 30°)$ **(ii)** $\cos135° = \cos(90° + 45°)$

**(iii)** $\tan15° = \tan(45° - 30°)$ **(iv)** $\tan75° = \tan(45° + 30°)$

**2** Expand each of the following expressions.

**(i)** $\sin(\theta + 45°)$ **(ii)** $\cos(\theta - 30°)$

**(iii)** $\sin(60° - \theta)$ **(iv)** $\cos(2\theta + 45°)$

**(v)** $\tan(\theta + 45°)$ **(vi)** $\tan(\theta - 45°)$

**3** Simplify each of the following expressions.

**(i)** $\sin 2\theta \cos\theta - \cos 2\theta \sin\theta$

**(ii)** $\cos\phi \cos 3\phi - \sin\phi \sin 3\phi$

**(iii)** $\sin 120° \cos 60° + \cos 120° \sin 60°$

**(iv)** $\cos\theta \cos\theta - \sin\theta \sin\theta$

**4** Solve the following equations for values of $\theta$ in the range $0° \leqslant \theta \leqslant 180°$.

**(i)** $\cos(60° + \theta) = \sin\theta$ **(ii)** $\sin(45° - \theta) = \cos\theta$

**(iii)** $\tan(45° + \theta) = \tan(45° - \theta)$ **(iv)** $2\sin\theta = 3\cos(\theta - 60°)$

**(v)** $\sin\theta = \cos(\theta + 120°)$

**5** Solve the following equations for values of $\theta$ in the range $0 \leqslant \theta \leqslant \pi$.
(When the range is given in radians, the solutions should be in radians, using multiples of $\pi$ where appropriate.)

**(i)** $\sin(\theta + \frac{\pi}{4}) = \cos\theta$

**(ii)** $2\cos(\theta - \frac{\pi}{3}) = \cos(\theta + \frac{\pi}{2})$

**6** Calculators are not to be used in this question.
The diagram shows three points L (–2, 1), M (0, 2) and N (3, –2) joined to form a triangle. The angles $\alpha$ and $\beta$ and the point P are shown in the diagram.

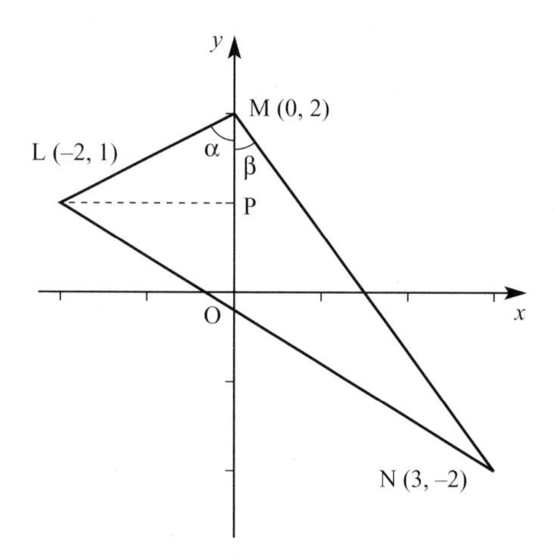


**(i)** Show that $\sin\alpha = \dfrac{2}{\sqrt{5}}$ and write down the value of $\cos\alpha$.

**(ii)** Find the values of $\sin\beta$ and $\cos\beta$.

**(iii)** Show that $\sin\angle\text{LMN} = \dfrac{11}{5\sqrt{5}}$.

**(iv)** Show that $\tan\angle\text{LNM} = \dfrac{11}{27}$.

[MEI]

# Double-angle formulae



Substituting $\phi = \theta$ in the relevant compound formulae leads immediately to expressions for $\sin 2\theta$, $\cos 2\theta$ and $\tan 2\theta$, as follows.

**(i)** $\sin(\theta + \phi) = \sin\theta\cos\phi + \cos\theta\sin\phi$

When $\phi = \theta$, this becomes

$$\sin(\theta + \theta) = \sin\theta\cos\theta + \cos\theta\sin\theta$$

giving $\sin 2\theta = 2\sin\theta\cos\theta$.

**(ii)** $\cos(\theta + \phi) = \cos\theta\cos\phi - \sin\theta\sin\phi$

When $\phi = \theta$, this becomes

$$\cos(\theta + \theta) = \cos\theta\cos\theta - \sin\theta\sin\theta$$

giving $\cos 2\theta = \cos^2\theta - \sin^2\theta$.

Using the Pythagorean identity $\cos^2\theta + \sin^2\theta = 1$, two other forms for $\cos 2\theta$ can be obtained:

$$\cos 2\theta = (1 - \sin^2\theta) - \sin^2\theta \quad \Rightarrow \quad \cos 2\theta = 1 - 2\sin^2\theta$$

$$\cos 2\theta = \cos^2\theta - (1 - \cos^2\theta) \quad \Rightarrow \quad \cos 2\theta = 2\cos^2\theta - 1.$$

These alternative forms are often more useful since they contain only one trigonometric function.

**(iii)** $\tan(\theta + \phi) = \dfrac{\tan\theta + \tan\phi}{1 - \tan\theta\tan\phi}$

When $\phi = \theta$ this becomes

$$\tan(\theta + \theta) = \frac{\tan\theta + \tan\theta}{1 - \tan\theta\tan\theta}$$

giving $\tan 2\theta = \dfrac{2\tan\theta}{1 - \tan^2\theta}$.

## Uses of the double-angle formulae

### IN MODELLING SITUATIONS

You will meet situations, such as that below, where using a double-angle formula not only allows you to write an expression more neatly but also thereby allows you to interpret its meaning more clearly.

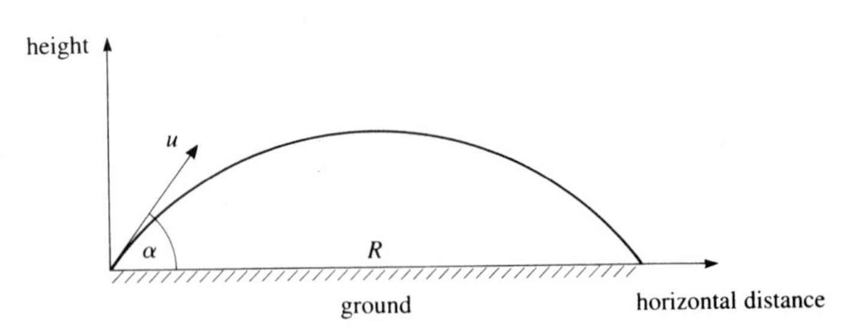


**Figure 2.3**

When an object is projected (e.g. a golf ball being hit as in figure 2.3) with speed $u$ at an angle $\alpha$ to the horizontal over level ground, the horizontal distance it travels before striking the ground, called its range, $R$, is given by the product of the horizontal component of the velocity $u\cos\alpha$ and its time of flight $\frac{(2u\sin\alpha)}{g}$:

$$R = \frac{2u^2\sin\alpha\cos\alpha}{g}.$$

Using the double-angle formula, $\sin2\alpha = 2\sin\alpha\cos\alpha$ allows this to be written as

$$R = \frac{u^2\sin2\alpha}{g}.$$

Since the maximum value of $\sin2\alpha$ is 1, it follows that the greatest value of the range $R$ is $\frac{u^2}{g}$ and that this occurs when $2\alpha = 90°$ and so $\alpha = 45°$. Thus an angle of projection of 45° will give the maximum range of the projectile over level ground. (This assumes that air resistance may be ignored.)

In this example, the double-angle formula enabled the expression for $R$ to be written tidily. However, it did more than that because it made it possible to find the maximum value of $R$ by inspection and without using calculus.

#### IN CALCULUS

The double-angle formulae allow a number of functions to be integrated and you will meet some of these in the next chapter.

The formulae for $\cos2\theta$ are particularly useful in this respect since

$$\cos2\theta = 1 - 2\sin^2\theta \quad\Rightarrow\quad \sin^2\theta = \tfrac{1}{2}(1 - \cos2\theta)$$

and

$$\cos2\theta = 2\cos^2\theta - 1 \quad\Rightarrow\quad \cos^2\theta = \tfrac{1}{2}(1 + \cos2\theta)$$

and these identities allow you to integrate $\sin^2\theta$ and $\cos^2\theta$.

### IN SOLVING EQUATIONS

You will sometimes need to solve equations involving both single and double angles as shown by the next two examples.

**EXAMPLE 2.2**

Solve the equation $\sin 2\theta = \sin\theta$ for $0° \leqslant \theta \leqslant 360°$.

**SOLUTION**

$$\sin 2\theta = \sin\theta$$

$$\Rightarrow \quad 2\sin\theta\cos\theta = \sin\theta$$

Be careful here: don't cancel by $\sin\theta$ or some roots will be lost

$$\Rightarrow \quad 2\sin\theta\cos\theta - \sin\theta = 0$$

$$\Rightarrow \quad \sin\theta(2\cos\theta - 1) = 0$$

$$\Rightarrow \quad \sin\theta = 0 \text{ or } \cos\theta = \tfrac{1}{2}.$$

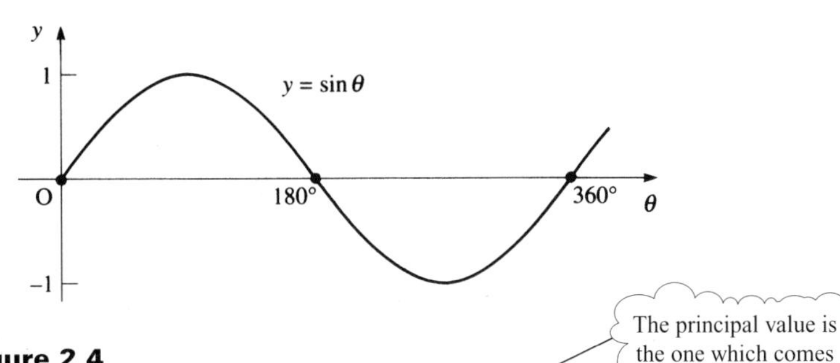


Figure 2.4

The principal value is the one which comes from your calculator

$\sin\theta = 0 \Rightarrow \theta = 0°$ (principal value) or 180° or 360° (see figure 2.4).

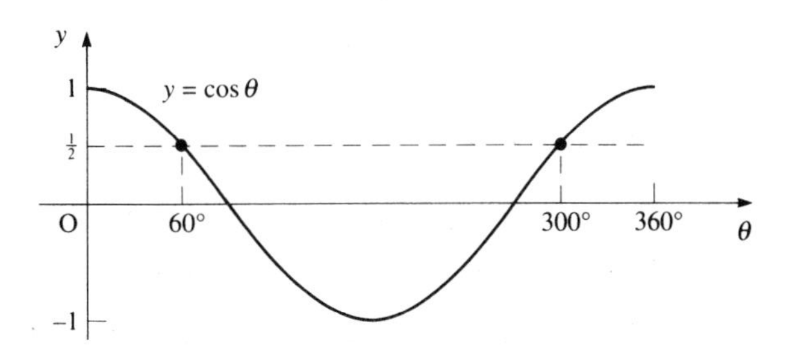


Figure 2.5

$\cos\theta = \frac{1}{2} \Rightarrow \theta = 60°$ (principal value) or 300° (see figure 2.5).

The full set of roots for $0 \leqslant \theta \leqslant 360°$ is $\theta = 0°, 60°, 180°, 300°, 360°$.

When an equation contains $\cos 2\theta$, you will save time if you take care to choose the most suitable expansion.

**EXAMPLE 2.3**

Solve $2 + \cos 2\theta = \sin\theta$ for $0 \leqslant \theta \leqslant 2\pi$. (Notice that the request for $\theta$ for $0 \leqslant \theta \leqslant 2\pi$, i.e. in radians, is an invitation to give the answer in radians.)

**SOLUTION**

This is the most suitable expansion since the right-hand side contains $\sin\theta$

Using $\cos 2\theta = 1 - 2\sin^2\theta$ gives

$$2 + (1 - 2\sin^2\theta) = \sin\theta$$

$$\Rightarrow \quad 2\sin^2\theta + \sin\theta - 3 = 0$$

$$\Rightarrow \quad (2\sin\theta + 3)(\sin\theta - 1) = 0$$

$\Rightarrow \quad \sin\theta = -\frac{3}{2}$ (not valid since $-1 \leqslant \sin\theta \leqslant 1$)

or $\quad \sin\theta = 1$.

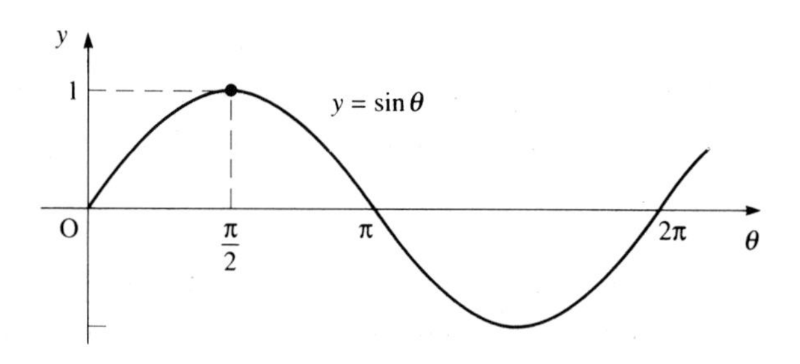


**Figure 2.6**

Figure 2.6 shows that the principal value $\theta = \frac{\pi}{2}$ is the only root for $0 \leqslant \theta \leqslant 2\pi$.

**EXERCISE 2B**

**1** Solve the following equations for $0° \leqslant \theta \leqslant 360°$.

**(i)** $2\sin 2\theta = \cos\theta$ **(ii)** $\tan 2\theta = 4\tan\theta$ **(iii)** $\cos 2\theta + \sin\theta = 0$

**(iv)** $\tan\theta\tan 2\theta = 1$ **(v)** $2\cos 2\theta = 1 + \cos\theta$

**2** Solve the following equations for $-\pi \leqslant \theta \leqslant \pi$.

**(i)** $\sin 2\theta = 2\sin\theta$ **(ii)** $\tan 2\theta = 2\tan\theta$ **(iii)** $\cos 2\theta - \cos\theta = 0$

**(iv)** $1 + \cos 2\theta = 2\sin^2\theta$ **(v)** $\sin 4\theta = \cos 2\theta$

(**Hint:** Express this as an equation in $2\theta$.)

**3** By first writing $\sin 3\theta$ as $\sin(2\theta + \theta)$, express $\sin 3\theta$ in terms of $\sin\theta$. Hence solve the equation $\sin 3\theta = \sin\theta$ for $0 \leqslant \theta \leqslant 2\pi$.

**4** Solve $\cos 3\theta = 1 - 3\cos\theta$ for $0° \leqslant \theta \leqslant 360°$.

**5** Simplify $\dfrac{1 + \cos 2\theta}{\sin 2\theta}$.

**6** Express $\tan 3\theta$ in terms of $\tan\theta$.

**7** Show that $\dfrac{1-\tan^2\theta}{1+\tan^2\theta} = \cos 2\theta$.

**8 (i)** Show that $\tan(\frac{\pi}{4}+\theta)\tan(\frac{\pi}{4}-\theta) = 1$.

**(ii)** Given that $\tan 26.6° = 0.5$, solve $\tan\theta = 2$ without using your calculator. Give $\theta$ to 1 decimal place, where $0° < \theta < 90°$.

**9 (i)** Sketch on the same axes the graphs of

$$y = \cos 2x \quad \text{and} \quad y = 3\sin x - 1 \quad \text{for} \quad 0 \leqslant x \leqslant 2\pi.$$

**(ii)** Show that these curves meet at points whose $x$ co-ordinates are solutions of the equation $2\sin^2 x + 3\sin x - 2 = 0$.

**(iii)** Solve this equation to find the values of $x$ in terms of $\pi$ for $0 \leqslant x \leqslant 2\pi$.

[MEI]

## The factor formulae

The factor formulae (pages 37 to 40) are not a requirement of the MEI Pure Mathematics 3 subject criteria. They are included here because they are very useful and are an application of the compound-angle formulae.

In algebra, the term ‘factorising’ means writing expressions as products. For example, ‘factorise $x^2 - 3x + 2$’ means ‘write $x^2 - 3x + 2$ as $(x-1)(x-2)$’. The same idea of factorising applies in trigonometry: you write sums or differences of trigonometric functions as products.

The factor formulae are derived from the compound-angle formulae.

Start with the compound-angle formulae for $\sin(\theta + \phi)$ and $\sin(\theta - \phi)$:

$$\sin(\theta + \phi) = \sin\theta\cos\phi + \cos\theta\sin\phi \qquad ①$$

$$\sin(\theta - \phi) = \sin\theta\cos\phi - \cos\theta\sin\phi. \qquad ②$$

Adding ① and ② gives

$$\sin(\theta + \phi) + \sin(\theta - \phi) = 2\sin\theta\cos\phi. \qquad ③$$

At this point, it is helpful to change variables by writing

$$\theta + \phi = \alpha \quad \text{and} \quad \theta - \phi = \beta$$

so that $\theta = \frac{1}{2}(\alpha + \beta)$ and $\phi = \frac{1}{2}(\alpha - \beta)$.

Substituting for $\theta$ and $\phi$ in ③ gives

$$\sin\alpha + \sin\beta = 2\sin\left(\frac{\alpha+\beta}{2}\right)\cos\left(\frac{\alpha-\beta}{2}\right).$$

The left-hand side is a sum, and the right-hand side is a product, so the expression has been factorised.

Similarly, subtracting ② from ① gives

$$\sin\alpha - \sin\beta = 2\cos\left(\frac{\alpha+\beta}{2}\right)\sin\left(\frac{\alpha-\beta}{2}\right).$$

ACTIVITY

Write down the expressions for $\cos(\theta + \phi)$ and $\cos(\theta - \phi)$ and use these to obtain factor formulae for $\cos\alpha + \cos\beta$ and for $\cos\alpha - \cos\beta$. Check your answers with the Key points at the end of the chapter.

## When do you use the factor formulae?

### ADDITION OF WAVEFORMS

The factor formulae allow you to add together sine and cosine functions. This operation is equivalent to the physical situation of combining waves of the same size (amplitude).

INVESTIGATION

### TWO MUSICIANS PLAYING IN TUNE

The sound of two musicians playing in tune with the same loudness may be modelled as two waves given by $x_1 = a\sin\omega t$ and $x_2 = a\sin(\omega t + \varepsilon)$.

The constant $\omega$ is related to the frequency of these waves and so to the pitch of the musical notes. (The frequency is given by $\frac{\omega}{2\pi}$). The two waves are not in phase and this is represented by the constant $\varepsilon$ in the expression for $x_2$ (see figure 2.7).

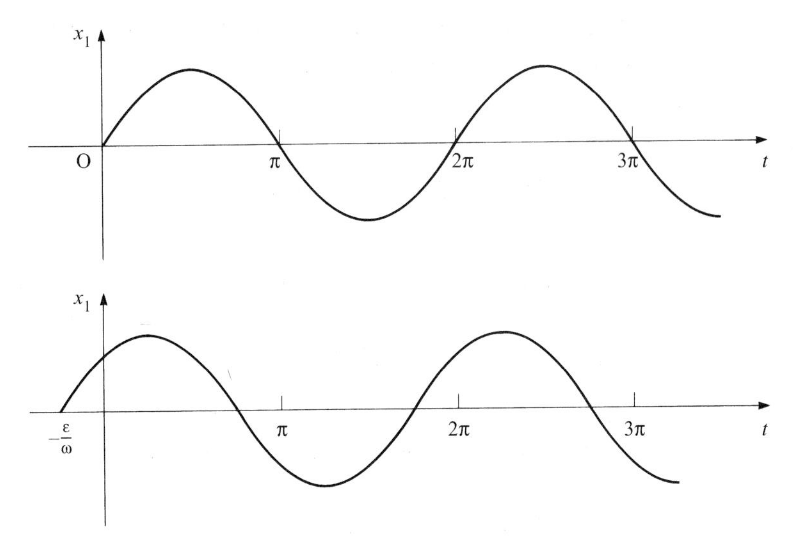


**Figure 2.7**

Show that $x_1 + x_2$ is a single wave. That is, the musicians sound as one but louder.

#### TWO MUSICIANS PLAYING SLIGHTLY OUT OF TUNE

In this case, the waves are given by $x_1 = a\sin\omega t$ and $x_2 = a\sin(\omega + \delta)t$, where $\delta$ is very small compared to $\omega$. Find the expression for $x_1 + x_2$.

Explain how this makes the combined note of the musicians vary in loudness, a phenomenon known as *beats*. How do beats help a piano tuner?

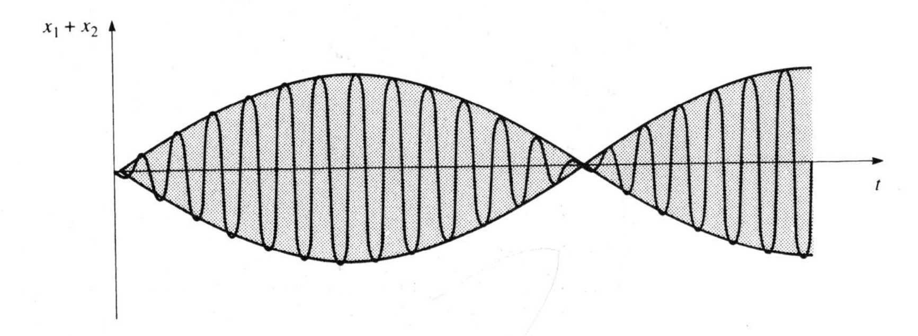


**Figure 2.8**

#### MANIPULATION

The factor formulae are often useful in tidying up expressions and in solving equations, as in the next example.

**EXAMPLE 2.4**

Solve $\sin 3\theta + \sin\theta = 0$ for $0° \leqslant \theta \leqslant 360°$.

**SOLUTION**

Using

$$\sin\alpha + \sin\beta = 2\sin\left(\frac{\alpha+\beta}{2}\right)\cos\left(\frac{\alpha-\beta}{2}\right)$$

and putting $\alpha = 3\theta$ and $\beta = \theta$ gives

$$\sin 3\theta + \sin\theta = 2\sin 2\theta\cos\theta$$

so the equation becomes

$$2\sin 2\theta\cos\theta = 0$$

$$\Rightarrow \quad \cos\theta = 0 \quad \text{or} \quad \sin 2\theta = 0.$$

From the graphs for $y = \cos\theta$ and $y = \sin\theta$

$\cos\theta = 0$ gives $\theta = 90°$ or $270°$

$\sin 2\theta = 0$ gives $2\theta = 0°, 180°, 360°, 540°$ or $720°$

so $\theta = 0°, 90°, 180°, 270°$ or $360°$.

You should only list each root once in the final answer

The complete set of roots in the range given is $\theta = 0°, 90°, 180°, 270°, 360°$.

**EXERCISE 2C**

**1** Factorise the following expressions.

**(i)** $\sin 4\theta - \sin 2\theta$

**(ii)** $\cos 5\theta + \cos\theta$

**(iii)** $\cos 7\theta - \cos 3\theta$

**(iv)** $\cos(\theta + 60°) + \cos(\theta - 60°)$

**(v)** $\sin(3\theta + 45°) + \sin(3\theta - 45°)$

**2** Factorise $\cos 4\theta + \cos 2\theta$. Hence, for $0° < \theta < 180°$ solve

$$\cos 4\theta + \cos 2\theta = \cos\theta.$$

**3** Simplify $\dfrac{\sin 5\theta + \sin 3\theta}{\sin 5\theta - \sin 3\theta}$.

**4** Solve the equation $\sin 3\theta - \sin\theta = 0$ for $0 \leqslant \theta \leqslant 2\pi$.

**5** Factorise $\sin(\theta + 73°) - \sin(\theta + 13°)$ and use your result to sketch the graph of $y = \sin(\theta + 73°) - \sin(\theta + 13°)$.

**6** Prove that $\sin^2 A - \sin^2 B = \sin(A - B)\sin(A + B)$.

**7** **(i)** Use a suitable factor formula to show that

$$\sin 3\theta + \sin\theta = 4\sin\theta\cos^2\theta.$$

**(ii)** Hence show that $\sin 3\theta = 3\sin\theta - 4\sin^3\theta$.

## The forms $r\cos(\theta \pm \alpha)$, $r\sin(\theta \pm \alpha)$

Another modification of the compound-angle formulae allows you to simplify expressions such as $4\sin\theta + 3\cos\theta$ and hence solve equations of the form

$$a\sin\theta + b\cos\theta = c.$$

To find a single expression for $4\sin\theta + 3\cos\theta$, you match it to the expression

$$r\sin(\theta + \alpha) = r(\sin\theta\cos\alpha + \cos\theta\sin\alpha).$$

This is because the expansion of $r\sin(\theta + \alpha)$ has $\sin\theta$ in the first term, $\cos\theta$ in the second term and a plus sign in between them. It is then possible to choose appropriate values of $r$ and $\alpha$:

$$4\sin\theta + 3\cos\theta \equiv r(\sin\theta\cos\alpha + \cos\theta\sin\alpha)$$

Coefficients of $\sin\theta$: $\quad 4 = r\cos\alpha$

Coefficients of $\cos\theta$: $\quad 3 = r\sin\alpha$.

Looking at the right-angled triangle in figure 2.9 gives the values for $r$ and $\alpha$.

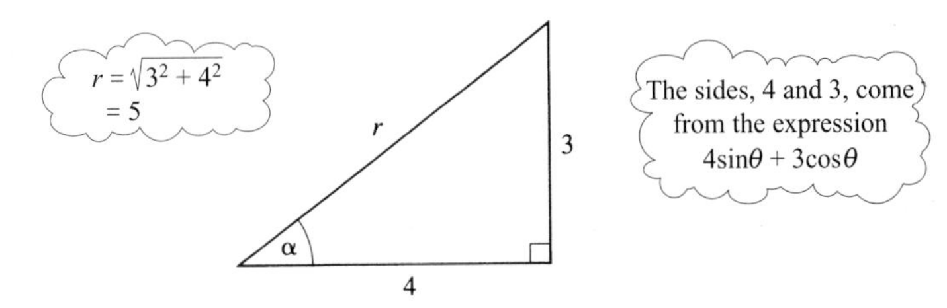


**Figure 2.9**

In this triangle, the hypotenuse is $\sqrt{4^2 + 3^2} = 5$, which corresponds to $r$ in the above expression.

The angle $\alpha$ is given by

$$\sin\alpha = \tfrac{3}{5} \quad \text{and} \quad \cos\alpha = \tfrac{4}{5} \quad \Rightarrow \quad \alpha = 36.9°.$$

So the expression becomes

$$4\sin\theta + 3\cos\theta = 5\sin(\theta + 36.9°).$$

The steps involved in this procedure can be generalised to write

$$a\sin\theta + b\cos\theta = r\sin(\theta + \alpha)$$

where

$$r = \sqrt{a^2 + b^2} \qquad \sin\alpha = \frac{b}{\sqrt{a^2 + b^2}} \qquad \cos\alpha = \frac{a}{\sqrt{a^2 + b^2}}.$$

The same expression may also be written as a cosine function. In this case, rewrite $4\sin\theta + 3\cos\theta$ as $3\cos\theta + 4\sin\theta$ and notice that:

**(i)** The expansion of $\cos(\theta - \beta)$ starts with $\cos\theta$ ... just like the expression $3\cos\theta + 4\sin\theta$.

**(ii)** The expansion of $\cos(\theta - \beta)$ has + in the middle, just like the expression $3\cos\theta + 4\sin\theta$.

The expansion of $r\cos(\theta - \beta)$ is given by

$$r\cos(\theta - \beta) = r(\cos\theta\cos\beta + \sin\theta\sin\beta).$$

To compare this with $3\cos\theta + 4\sin\theta$, look at the triangle (figure 2.10) in which

$$r = \sqrt{3^2 + 4^2} = 5 \qquad \cos\beta = \tfrac{3}{5} \qquad \sin\beta = \tfrac{4}{5} \qquad \Rightarrow \qquad \beta = 53.1°.$$

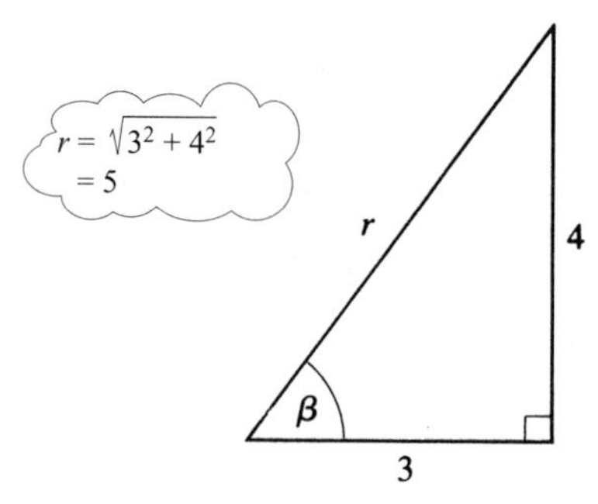


**Figure 2.10**

This means that you can write $3\cos\theta + 4\sin\theta$ in the form

$$r\cos(\theta - \beta) = 5\cos(\theta - 53.1°).$$

The procedure used here can be generalised to give the result

$$a\cos\theta + b\sin\theta = r\cos(\theta - \alpha)$$

where $r = \sqrt{a^2 + b^2} \qquad \cos\alpha = \dfrac{a}{r} \qquad \sin\alpha = \dfrac{b}{r}.$

***Note***

The value of $r$ will always be positive, but $\cos\alpha$ and $\sin\alpha$ may be positive or negative, depending on the values of $a$ and $b$. In all cases, it is possible to find an angle $\alpha$ for which $-180° < \alpha < 180°$.

You can derive alternative expressions of this type based on other compound-angle formulae if you wish $\alpha$ to be an acute angle, as is done in the next example.

**EXAMPLE 2.5**

**(i)** Express $\sqrt{3}\sin\theta - \cos\theta$ in the form $r\sin(\theta - \alpha)$, where $r > 0$ and $0 < \alpha < \frac{\pi}{2}$.

**(ii)** State the maximum and minimum values of $\sqrt{3}\sin\theta - \cos\theta$.

**(iii)** Sketch the graph of $y = \sqrt{3}\sin\theta - \cos\theta$ for $0 \leqslant \theta \leqslant 2\pi$.

**(iv)** Solve the equation $\sqrt{3}\sin\theta - \cos\theta = 1$ for $0 \leqslant \theta \leqslant 2\pi$.

**SOLUTION**

**(i)** $r\sin(\theta - \alpha) = r(\sin\theta\cos\alpha - \cos\theta\sin\alpha)$

$$= (r\cos\alpha)\sin\theta - (r\sin\alpha)\cos\theta.$$

Comparing this with $\sqrt{3}\sin\theta - \cos\theta$, the two expressions are identical if

$$r\cos\alpha = \sqrt{3} \qquad \text{and} \qquad r\sin\alpha = 1.$$

From the triangle in figure 2.11

$$r = \sqrt{1+3} = 2 \quad \text{and} \quad \tan\alpha = \frac{1}{\sqrt{3}} \Rightarrow \alpha = \tfrac{\pi}{6}$$

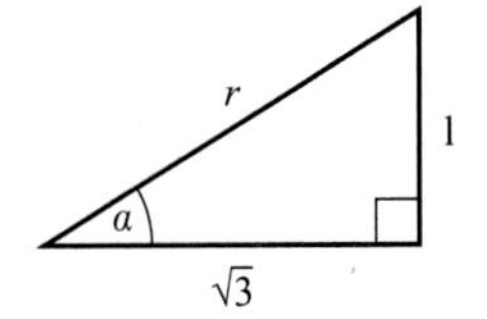


**Figure 2.11**

so $\qquad \sqrt{3}\sin\theta - \cos\theta = 2\sin(\theta - \tfrac{\pi}{6}).$

**(ii)** The sine function oscillates between 1 and –1, so $2\sin(\theta - \frac{\pi}{6})$ oscillates between 2 and –2.

Maximum value = 2

Minimum value = –2.

**(iii)** The graph of $y = 2\sin(\theta - \frac{\pi}{6})$ in figure 2.12 is obtained from the graph of $y = \sin\theta$ by a translation $\begin{pmatrix} \frac{\pi}{6} \\ 0 \end{pmatrix}$ and a stretch of factor 2 parallel to the $y$ axis.

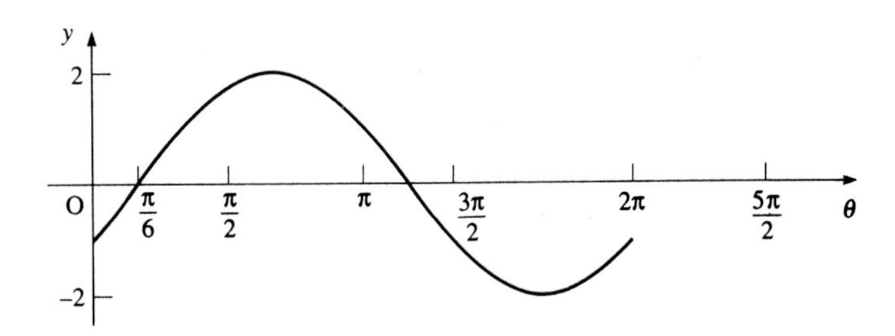


**Figure 2.12**

**(iv)** The equation $\sqrt{3}\sin\theta - \cos\theta = 1$ is equivalent to

$$2\sin(\theta - \tfrac{\pi}{6}) = 1$$

$$\Rightarrow \quad \sin(\theta - \tfrac{\pi}{6}) = \tfrac{1}{2}.$$

Let $x = (\theta - \frac{\pi}{6})$ and solve $\sin x = \frac{1}{2}$.

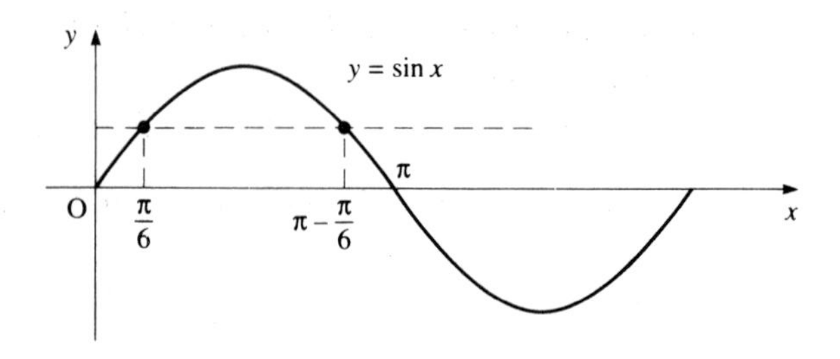


**Figure 2.13**

Solving $\sin x = \frac{1}{2}$ gives $x = \frac{\pi}{6}$ (principal value) or

$x = \pi - \frac{\pi}{6} = \frac{5\pi}{6}$ (from the graph in figure 2.13)

giving $\theta = \frac{\pi}{6} + \frac{\pi}{6} = \frac{\pi}{3}$ or $\theta = \frac{5\pi}{6} + \frac{\pi}{6} = \pi$.

The roots in $0 \leqslant \theta \leqslant 2\pi$ are $\theta = \frac{\pi}{3}$ and $\pi$.

Always check (e.g. by reference to a sketch graph) that the number of roots you have found is consistent with the number you are expecting. When solving equations of the form $\sin(\theta - \alpha) = c$ by considering $\sin x = c$, it is sometimes necessary to go outside the range specified for $\theta$ since, for example, $0 \leqslant \theta \leqslant 2\pi$ is the same as $-\alpha \leqslant x \leqslant 2\pi - \alpha$.

### Using these forms

There are many situations, as on page 41, which produce expressions which can be tidied up using these forms. They are also particularly useful for solving equations involving both the sine and cosine of the same angle.

The fact that $a\cos\theta + b\sin\theta$ can be written as $r\cos(\theta - \alpha)$ is an illustration of the fact that any two waves of the same frequency, whatever their amplitudes, can be added together to give a single combined wave, also of the same frequency.

**EXERCISE 2D**

**1** Express each of the following in the form $r\cos(\theta - \alpha)$, where $r > 0$ and $0 < \alpha < 90°$.

**(i)** $\cos\theta + \sin\theta$ **(ii)** $3\cos\theta + 4\sin\theta$

**(iii)** $\cos\theta + \sqrt{3}\sin\theta$ **(iv)** $\sqrt{5}\cos\theta + 2\sin\theta$

**2** Express each of the following in the form $r\cos(\theta + \alpha)$, where $r > 0$ and $0 < \alpha < \frac{\pi}{2}$.

**(i)** $\cos\theta - \sin\theta$ **(ii)** $\sqrt{3}\cos\theta - \sin\theta$

**3** Express each of the following in the form $r\sin(\theta + \alpha)$, where $r > 0$ and $0 < \alpha < 90°$.

**(i)** $\sin\theta + 2\cos\theta$ **(ii)** $3\sin\theta + 4\cos\theta$

**4** Express each of the following in the form $r\sin(\theta - \alpha)$, where $r > 0$ and $0 < \alpha < \frac{\pi}{2}$.

**(i)** $\sin\theta - \cos\theta$ **(ii)** $\sqrt{3}\sin\theta - \cos\theta$

**5** Express each of the following in the form $r\cos(\theta - \alpha)$, where $r > 0$ and $-180° < \alpha < 180°$.

**(i)** $\cos\theta - \sqrt{3}\sin\theta$ **(ii)** $2\sqrt{2}\cos\theta - 2\sqrt{2}\sin\theta$

**(iii)** $\sin\theta + \sqrt{3}\cos\theta$ **(iv)** $5\sin\theta + 12\cos\theta$

**(v)** $\sin\theta - \sqrt{3}\cos\theta$ **(vi)** $\sqrt{2}\sin\theta - \sqrt{2}\cos\theta$

**6** **(i)** Express $5\cos\theta - 12\sin\theta$ in the form $r\cos(\theta + \alpha)$, where $r > 0$ and $0 < \alpha < 90°$.

**(ii)** State the maximum and minimum values of $5\cos\theta - 12\sin\theta$.

**(iii)** Sketch the graph of $y = 5\cos\theta - 12\sin\theta$ for $0 \leqslant \theta \leqslant 360°$.

**(iv)** Solve the equation $5\cos\theta - 12\sin\theta = 4$ for $0 \leqslant \theta \leqslant 360°$.

**7** **(i)** Express $3\sin\theta - \sqrt{3}\cos\theta$ in the form $r\sin(\theta - \alpha)$, where $r > 0$ and $0 < \alpha < \frac{\pi}{2}$.

**(ii)** State the maximum and minimum values of $3\sin\theta - \sqrt{3}\cos\theta$ and the smallest positive values of $\theta$ for which they occur.

**(iii)** Sketch the graph of $y = 3\sin\theta - \sqrt{3}\cos\theta$ for $0 \leqslant \theta \leqslant 2\pi$.

**(iv)** Solve the equation $3\sin\theta - \sqrt{3}\cos\theta = \sqrt{3}$ for $0 \leqslant \theta \leqslant 2\pi$.

**8** **(i)** Express $2\sin2\theta + 3\cos2\theta$ in the form $r\sin(2\theta + \alpha)$, where $r > 0$ and $0 < \alpha < 90°$.

**(ii)** State the maximum and minimum values of $2\sin2\theta + 3\cos2\theta$ and the smallest positive values of $\theta$ for which they occur.

**(iii)** Sketch the graph of $y = 2\sin2\theta + 3\cos2\theta$ for $0 \leqslant \theta \leqslant 360°$.

**(iv)** Solve the equation $2\sin2\theta + 3\cos2\theta = 1$ for $0 \leqslant \theta \leqslant 360°$.

**9** **(i)** Express $\cos\theta + \sqrt{2}\sin\theta$ in the form $r\cos(\theta - \alpha)$, where $r > 0$ and $0 < \alpha < 90°$.

**(ii)** State the maximum and minimum values of $\cos\theta + \sqrt{2}\sin\theta$ and the smallest positive values of $\theta$ for which they occur.

**(iii)** Sketch the graph of $y = \cos\theta + \sqrt{2}\sin\theta$ for $0 \leqslant \theta \leqslant 360°$.

**(iv)** State the maximum and minimum values of

$$\frac{1}{3 + \cos\theta + \sqrt{2}\sin\theta}$$

and the smallest positive values for which they occur.

**10** The diagram shows a table jammed in a corridor. The table is 120 cm long and 80 cm wide, and the width of the corridor is 130 cm.

**(i)** Show that $12\sin\theta + 8\cos\theta = 13$.

**(ii)** Hence find the angle $\theta$. (There are two answers.)

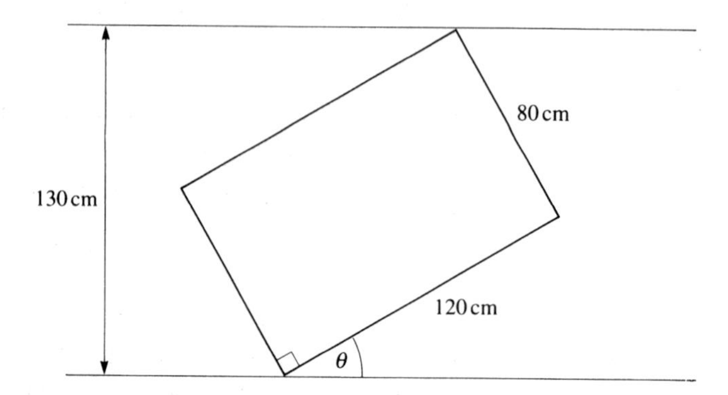


**11 (i)** Use a trigonometrical formula to expand $\cos(x + \alpha)$.

**(ii)** Express $y = 2\cos x - 5\sin x$ in the form $r\cos(x + \alpha)$, giving the positive value of $r$ and the smallest positive value of $\alpha$.

**(iii)** State the maximum and minimum values of $y$ and the corresponding values of $x$ for $0° \leqslant x \leqslant 360°$.

**(iv)** Solve the equation

$$2\cos x - 5\sin x = 3, \quad \text{for } 0° \leqslant x \leqslant 360°.$$

[MEI]

**12 (i)** Find the value of the acute angle $\alpha$ for which

$$5\cos x - 3\sin x = \sqrt{34}\cos(x + \alpha)$$

for all $x$.

Giving your answers correct to 1 decimal place,

**(ii)** solve the equation $5\cos x - 3\sin x = 4$ for $0° \leqslant x \leqslant 360°$,

**(iii)** solve the equation $5\cos 2x - 3\sin 2x = 4$ for $0° \leqslant x \leqslant 360°$.

[MEI]

**13 (i)** Find the positive value of $R$ and the acute angle $\alpha$ for which

$$6\cos x + 8\sin x = R\cos(x - \alpha).$$

**(ii)** Sketch the curve with equation

$$y = 6\cos x + 8\sin x, \quad \text{for } 0° \leqslant x \leqslant 360°.$$

Mark your axes carefully and indicate the angle $\alpha$ on the $x$ axis.

**(iii)** Solve the equation

$$6\cos x + 8\sin x = 4, \quad \text{for } 0° \leqslant x \leqslant 360°.$$

**(iv)** Solve the equation

$$8\cos\theta + 6\sin\theta = 4, \quad \text{for } 0° \leqslant \theta \leqslant 360°.$$

[MEI]

**14 (i)** Express $3\cos x + \sin x$ in the form $R\cos(x - \alpha)$, giving the value of $R$ and the smallest positive value of $\alpha$.

**(ii)** Use your answer to part (i) to solve the equation

$$3\cos x + \sin x = 1, \quad \text{for } 0° \leqslant x \leqslant 360°.$$

**(iii)** Solve the equation $(3\cos x)^2 = (1 - \sin x)^2$ by substituting for $\cos^2 x$ in terms of $\sin x$ and solving the resulting quadratic equation in $\sin x$.

**(iv)** Explain why the answers to (ii) and (iii) are not the same.

[MEI]

**15** In the diagram below, angle QPT = angle SQR = $\theta$, angle QPR = $\alpha$, PQ = $a$, QR = $b$, PR = $c$, angle QSR = angle QTP = 90°, SR = TU.

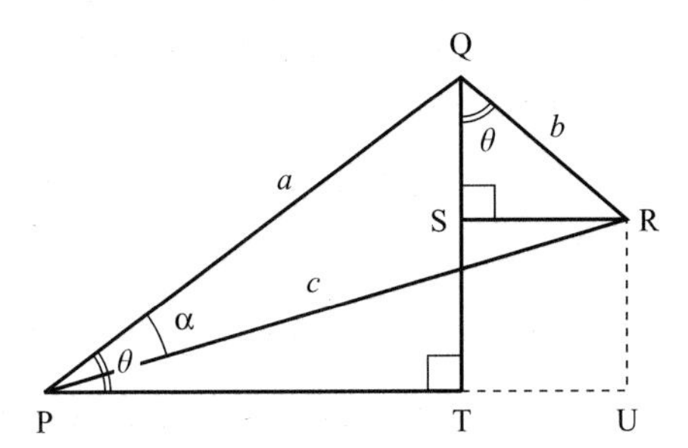


**(i)** Show that angle PQR = 90°, and write down the length of $c$ in terms of $a$ and $b$.

**(ii)** Show that PU may be written as $a\cos\theta + b\sin\theta$ and as $c\cos(\theta - \alpha)$. Write down the value of $\tan\alpha$ in terms of $a$ and $b$.

**(iii)** In the case when $a = 4$, $b = 3$, find the acute angle $\alpha$.

**(iv)** Solve the equation

$$4\cos\theta + 3\sin\theta = 2 \quad \text{for} \quad 0° \leqslant \theta \leqslant 360°.$$

[MEI]

## INVESTIGATION

The simplest alternating current is one which varies with time $t$ according to $I = A\sin 2\pi ft$, where $f$ is the frequency and $A$ is the maximum value. The frequency of the public AC supply is 50 hertz (cycles per second).

Investigate what happens when two alternating currents $A_1\sin 2\pi ft$ and $A_2\sin(2\pi ft + \alpha)$ with the same frequency $f$ but a phase difference of $\alpha$ are added together.

The previous exercises have each concentrated on just one of the many trigonometrical techniques which you will need to apply confidently. The following exercise requires you to identify which technique is the correct one.

## EXERCISE 2E

**1** Simplify:

**(i)** $2\sin 3\theta\cos 3\theta$

**(ii)** $\cos^2 3\theta - \sin^2 3\theta$

**(iii)** $\cos^2 3\theta + \sin^2 3\theta$

**(iv)** $1 - 2\sin^2\dfrac{\theta}{2}$

**(v)** $\sin(\theta - \alpha)\cos\alpha + \cos(\theta - \alpha)\sin\alpha$

**(vi)** $3\sin\theta\cos\theta$

**(vii)** $\dfrac{\sin 2\theta}{2\sin\theta}$

**(viii)** $\cos 2\theta - 2\cos^2\theta$.

**2** Express:

**(i)** $(\cos x - \sin x)^2$ in terms of $\sin 2x$

**(ii)** $\cos^4 x - \sin^4 x$ in terms of $\cos 2x$

**(iii)** $2\cos^2 x - 3\sin^2 x$ in terms of $\cos 2x$.

**3** Prove that:

**(i)** $\dfrac{1 - \cos 2\theta}{1 + \cos 2\theta} \equiv \tan^2\theta$

**(ii)** $\operatorname{cosec} 2\theta + \cot 2\theta \equiv \cot\theta$

**(iii)** $\tan 4\theta \equiv \dfrac{4t(1 - t^2)}{1 - 6t^2 + t^4}$ where $t = \tan\theta$.

4 Solve the following equations.

(i) $\sin(\theta + 40°) = 0.7$ $\quad 0° \leqslant \theta \leqslant 360°$

(ii) $3\cos^2\theta + 5\sin\theta - 1 = 0$ $\quad 0° \leqslant \theta \leqslant 360°$

(iii) $2\cos(\theta - \frac{\pi}{6}) = 1$ $\quad -\pi \leqslant \theta \leqslant \pi$

(iv) $\cos(45° - \theta) = 2\sin(30° + \theta)$ $\quad -180° \leqslant \theta \leqslant 180°$

(v) $\cos 2\theta + 3\sin\theta = 2$ $\quad 0 \leqslant \theta \leqslant 2\pi$

(vi) $\cos\theta + 3\sin\theta = 2$ $\quad 0 \leqslant \theta \leqslant 360°$

(vii) $\tan^2 x - 3\tan x - 4 = 0$ $\quad 0 \leqslant \theta \leqslant 180°$



## Small-angle approximations

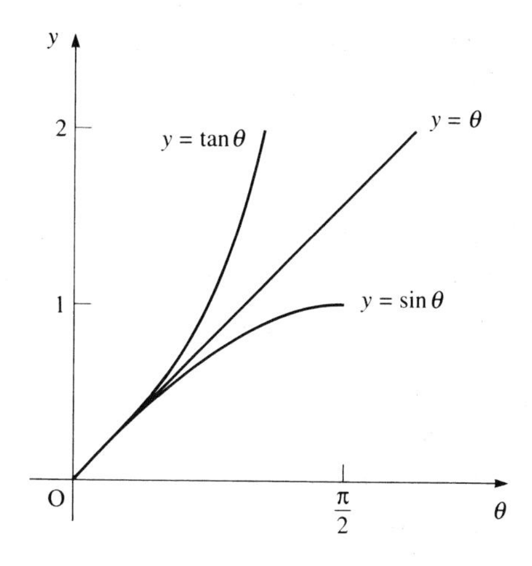


**Figure 2.14**

Figure 2.14 shows the graphs of $y = \theta$, $y = \sin\theta$ and $y = \tan\theta$ on the same axes, for $0 \leqslant \theta \leqslant \frac{\pi}{2}$. The same scale is used for both axes.

From this, it appears that in this interval, $\sin\theta < \theta < \tan\theta$.

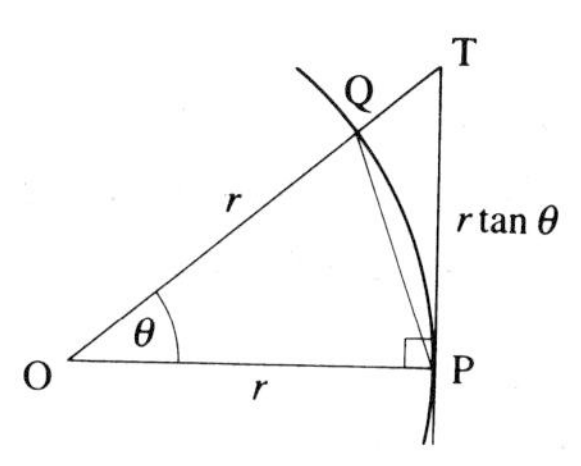


**Figure 2.15**

To prove this result, look at figure 2.15. PT is a tangent to the circle, radius $r$ units and centre O.

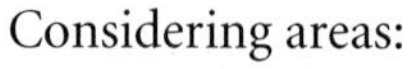

Considering areas:

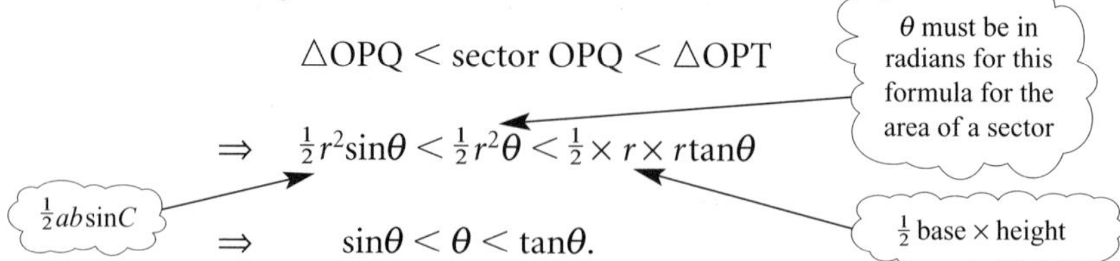


$$\triangle OPQ < \text{sector } OPQ < \triangle OPT$$

$$\Rightarrow \quad \tfrac{1}{2}r^2\sin\theta < \tfrac{1}{2}r^2\theta < \tfrac{1}{2} \times r \times r\tan\theta$$

$$\Rightarrow \quad \sin\theta < \theta < \tan\theta.$$

Use a graphics calculator to draw $y = \theta$, $y = \sin\theta$ and $y = \tan\theta$ on the same axes, for $0 \leqslant \theta \leqslant 0.2$ radians. Notice how close the graphs are. This suggests that for small values of $\theta$, $\sin\theta \approx \theta$ and $\tan\theta \approx \theta$.

The result $\sin\theta \approx \theta$ for small angles $\theta$ is a fundamental result which you will meet again in Chapter 3 when you differentiate trigonometric functions. To prove this, take the relationship $\sin\theta < \theta < \tan\theta$ proved earlier for $0 < \theta < \frac{\pi}{2}$ and divide through by $\sin\theta$ to give

$$1 < \frac{\theta}{\sin\theta} < \frac{\tan\theta}{\sin\theta}$$

$$\Rightarrow \quad 1 < \frac{\theta}{\sin\theta} < \frac{1}{\cos\theta}.$$

As $\theta \to 0$, $\cos\theta \to 1$, so $\dfrac{\theta}{\sin\theta}$ is sandwiched between 1 and something approaching 1, showing that as $\theta \to 0$, $\sin\theta \approx \theta$. This can be written formally as

$$\lim_{\theta \to 0} \frac{\theta}{\sin\theta} = 1$$

Dividing each term in the relationship $\sin\theta < \theta < \tan\theta$ by $\tan\theta$ gives

$$\frac{\sin\theta}{\tan\theta} < \frac{\theta}{\tan\theta} < 1$$

$$\Rightarrow \quad \cos\theta < \frac{\theta}{\tan\theta} < 1.$$

As $\theta \to 0$, $\cos\theta \to 1$, showing that $\lim\limits_{\theta \to 0} \dfrac{\theta}{\tan\theta} = 1$.

You know that $\cos 0 = 1$, and for small values of $\theta$, $\cos\theta \approx 1$ but it is easy to obtain a closer approximation.

Using the double-angle formula $\cos 2\theta = 1 - 2\sin^2\theta$ and replacing $2\theta$ by $\theta$ (and $\theta$ by $\frac{\theta}{2}$) gives

$$\cos\theta = 1 - 2\sin^2\frac{\theta}{2}. \qquad ①$$

When $\theta$ is small, so is $\frac{\theta}{2}$, so $\sin\frac{\theta}{2} \approx \frac{\theta}{2}$. In ① this gives

$$\cos\theta \approx 1 - 2\left(\frac{\theta}{2}\right)^2$$

$$\Rightarrow \quad \cos\theta \approx 1 - \frac{\theta^2}{2}.$$

All of these approximations are very good for $-0.1 \leqslant \theta \leqslant 0.1$ radians.

---

**?** What do you think is meant by the expression 'very good' above? Can you quantify it by calculating the maximum percentage error?

---

**EXAMPLE 2.6**

Given that $\theta$ is small, show that $\tan(\frac{\pi}{4} + \theta) \approx \frac{1+\theta}{1-\theta}$.

**SOLUTION**

$$\tan(\tfrac{\pi}{4} + \theta) = \frac{\tan\frac{\pi}{4} + \tan\theta}{1 - \tan\frac{\pi}{4}\tan\theta}$$

$$= \frac{1 + \tan\theta}{1 - \tan\theta} \qquad \text{since } \tan\tfrac{\pi}{4} = 1$$

$$\approx \frac{1+\theta}{1-\theta}.$$

These approximations can also be used to find the limit of a fractional expression as $\theta \to 0$ in cases when substituting $\theta = 0$ gives $\frac{0}{0}$, which is undefined.

**ACTIVITY**

**(i)** Show that substituting $\theta = 0$ into the expression

$$\frac{\cos\theta - \cos 2\theta}{\theta^2}$$

gives $\frac{0}{0}$, which is undefined.

**(ii)** Investigate the behaviour of this expression as $\theta \to 0$ by evaluating

$$\frac{\cos\theta - \cos 2\theta}{\theta^2}$$

for values of $\theta$ (in radians) starting with $\theta = 0.2$ and decreasing in steps of 0.02.

**EXAMPLE 2.7**

**(i)** Find an approximation for $\cos\theta - \cos2\theta$ when $\theta$ is small.

**(ii)** Hence find

$$\lim_{\theta \to 0} \frac{\cos\theta - \cos2\theta}{\theta^2}.$$

**SOLUTION**

**(i)** When $\theta$ and $2\theta$ are both small

$$\cos\theta \approx 1 - \frac{\theta^2}{2} \quad \text{and} \quad \cos2\theta \approx 1 - \frac{(2\theta)^2}{2}$$

$$\approx 1 - 2\theta^2.$$

Using these approximations, when $\theta$ is small

$$\cos\theta - \cos2\theta \approx \left(1 - \frac{\theta^2}{2}\right) - (1 - 2\theta^2)$$

$$= \frac{3\theta^2}{2}.$$

**(ii)**

$$\frac{\cos\theta - \cos2\theta}{\theta^2} \approx \frac{3\theta^2}{2\theta^2}$$

$$\approx \frac{3}{2}.$$

This is consistent with the result in the activity above and may be written as

$$\lim_{\theta \to 0} \frac{\cos\theta - \cos2\theta}{\theta^2} = \frac{3}{2}.$$

**EXAMPLE 2.8**

**(i)** Simplify $\tan(\frac{\pi}{4} + \theta)$ when $\theta$ is small.

**(ii)** Hence use the binomial theorem to find a quadratic approximation for $\tan(\frac{\pi}{4} + \theta)$ when $\theta$ is small.

**SOLUTION**

**(i)**

$$\tan(\tfrac{\pi}{4} + \theta) = \frac{\tan\frac{\pi}{4} + \tan\theta}{1 - \tan\frac{\pi}{4}\tan\theta}$$

$$= \frac{1 + \tan\theta}{1 - \tan\theta}$$

$$\approx \frac{1 + \theta}{1 - \theta} \quad \text{when } \theta \text{ is small.}$$

**(ii)** $$\frac{1+\theta}{1-\theta} = (1+\theta)(1-\theta)^{-1}$$

$$= (1+\theta)\left[1 + (-1)(-\theta) + \frac{(-1)(-2)(-\theta)^2}{2!} + \dots\right]$$

$$\approx (1+\theta)(1+\theta+\theta^2)$$

$$\approx 1 + 2\theta + 2\theta^2.$$

## EXERCISE 2F

**1** When $\theta$ is small enough for $\theta^3$ to be ignored, find approximate expressions for the following.

**(i)** $\dfrac{\theta\sin\theta}{1-\cos\theta}$ **(ii)** $2\cos(\frac{\pi}{3}+\theta)$

**(iii)** $\cos\theta\cos 2\theta$ **(iv)** $\dfrac{\theta\tan\theta}{1-\cos 2\theta}$

**(v)** $\dfrac{\cos 4\theta - \cos 2\theta}{\sin 4\theta - \sin 2\theta}$ **(vi)** $\sin(\alpha+\theta)\sin\theta$ (Note: $\alpha$ is not small.)

**2** **(i)** Find an approximate expression for $\sin 2\theta + \tan 3\theta$ when $\theta$ is small enough for $3\theta$ to be considered as small.

**(ii)** Hence find

$$\lim_{\theta \to 0} \frac{\sin 2\theta + \tan 3\theta}{\theta}.$$

**3** **(i)** Find an approximate expression for $1 - \cos\theta$ when $\theta$ is small.

**(ii)** Hence find

$$\lim_{\theta \to 0} \frac{1-\cos\theta}{4\theta\sin\theta}.$$

**4** **(i)** Find an approximate expression for $\sin\theta[\sin(\frac{\pi}{6}+\theta) - \sin\frac{\pi}{6}]$ when $\theta$ is small.

**(ii)** Find an approximate expression for $1 - \cos 2\theta$ when $\theta$ is small.

**(iii)** Hence find

$$\lim_{\theta \to 0} \frac{\sin\theta[\sin(\frac{\pi}{6}+\theta) - \sin\frac{\pi}{6}]}{1-\cos 2\theta}.$$

**5** **(i)** Find an approximate expression for $1 - \cos 4\theta$ when $\theta$ is small enough for $4\theta$ to be considered as small.

**(ii)** Find an approximate expression for $\tan^2 2\theta$ when $\theta$ is small enough for $2\theta$ to be considered as small.

**(iii)** Hence find

$$\lim_{\theta \to 0} \frac{1-\cos 4\theta}{\tan^2 2\theta}.$$

**6** **(i)** Find an approximate expression for $\dfrac{1}{1+\tan\theta}$ when $\theta$ is small.

**(ii)** Hence use the binomial theorem to find a quadratic approximation for $\dfrac{1}{1+\tan\theta}$.

**(iii)** When $\theta = 0.1$ radians, find the percentage errors which arise when you use each of the expressions you have derived in parts (i) and (ii) in place of $\dfrac{1}{1+\tan\theta}$.

**7** **(i)** Find an approximate expression for $\sqrt{1+\sin\theta}$ when $\theta$ is small.

**(ii)** Hence use the binomial theorem to find a quadratic approximation for $\sqrt{1+\sin\theta}$.

**(iii)** Say which of these approximations you would expect to be the more accurate, and give a reason for your answer.

**(iv)** Check your answer to part (iii) by substituting $\theta = 0.1$ radians.

**8** **(i)** By writing $\sec\theta$ as $\dfrac{1}{\cos\theta}$ find an approximate expression for $\sec\theta$ when $\theta$ is small.

**(ii)** Hence use the binomial theorem to find a quadratic approximation for $\sec\theta$.

**(iii)** Use a trial and improvement method to find the largest value of $\theta$ for which the error incurred in using your answer to part (ii) in place of $\sec\theta$ is less than 1%.

**(iv)** Comment on your answer to part (iii).

**9** There are regulations in fencing to ensure that the blades used are not too bent. For épées, the rule states that the blade must not depart by more than 1 cm from the straight line joining the base to the point (figure A). For sabres, the corresponding rule states that the point must not be more than 4 cm out of line, i.e. away from the tangent at the base of the blade (figure B).

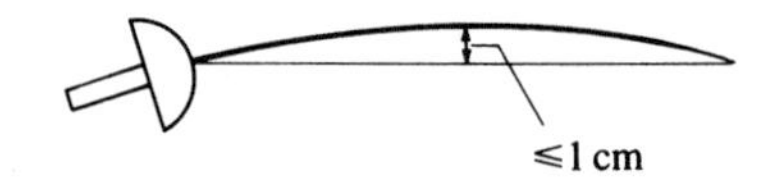


**Figure A**

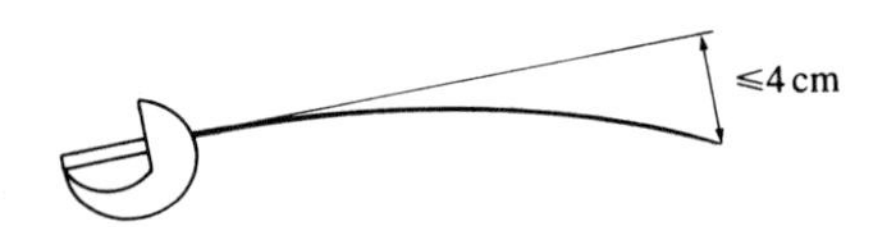


**Figure B**

Suppose that a blade AB is bent to form an arc of a circle of radius $r$, and that AB subtends an angle $2\theta$ at the centre O of the circle. Then with the notation of figure C, the épée bend is measured by CD, and the sabre bend by BE.

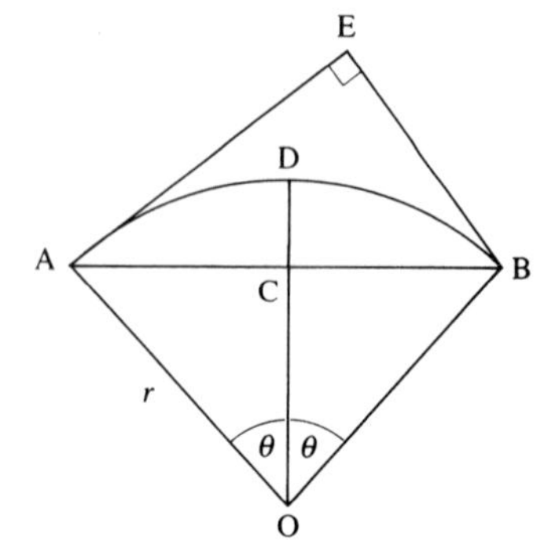


**Figure C**

**(i)** Show that CD = $r(1 - \cos\theta)$.

**(ii)** Explain why angle BAE = $\theta$.

**(iii)** Show that BE = $2r\sin^2\theta$.

**(iv)** Deduce that if $\theta$ is small, BE $\approx$ 4 CD and hence that the rules for épée and sabre amount to the same thing.

**10 (i)** Solve the equation $\sin 2x + \cos 2x = 0$ for $-\frac{\pi}{2} \leqslant x \leqslant \frac{\pi}{2}$, giving your answers in radians.

**(ii)** Show that $\cos 2x \approx 1 - 2x^2$ for small values of $x$. Write down a small-angle approximation for $\sin 2x$.

**(iii)** Using the results in (ii), find a quadratic function Q($x$) which is an approximation to $\sin 2x + \cos 2x$ for small values of $x$.

**(iv)** Solve the equation Q($x$) = 0.

**(v)** Comment on your answers to parts (i) and (iv).

[MEI]

**11 (i)** Express $\dfrac{2}{(2-x)(1-x)}$ in partial fractions.

Show that, for small values of $x$

$$\frac{2}{(2-x)(1-x)} \approx 1 + kx + \tfrac{7}{4}x^2$$

where $k$ is to be found.

**(ii)** By using a suitable small-angle approximation for $\cos\theta$, together with the result of part (i), show that, for small values of $\theta$

$$\frac{2}{(1+\cos\theta)\cos\theta} \approx 1 + \tfrac{3}{4}\theta^2.$$

**(iii)** Given that $\theta$ is small, find an approximate solution of the equation

$$\frac{2}{(1+\cos\theta)\cos\theta} = 0.99 + \sin^2\theta.$$

[MEI]

**12 (i)** Use a compound-angle formula to write down an expression for $\sin(x + \delta x)$.

**(ii)** Rewrite your answer to part (i) using small-angle approximations for $\sin\delta x$ and $\cos\delta x$.

**(iii)** Use your answer to part (ii) to write down an expression for

$$\frac{\sin(x+\delta x) - \sin x}{\delta x}.$$

**(iv)** State

$$\lim_{\delta x \to 0} \frac{\sin(x+\delta x) - \sin x}{\delta x}.$$

**(v)** Explain the significance of your answer to part (iv).

**INVESTIGATION**

Explain why tan89° is approximately but not exactly equal to the number of degrees in a radian.

## The general solutions of trigonometric equations

The equation $\tan\theta = 1$ has infinitely many roots:

$$\ldots, -315°, \quad -135°, \quad 45°, \quad 225°, \quad 405°, \ldots \text{ (in degrees)}$$

$$\ldots, -\frac{7\pi}{4}, \quad -\frac{3\pi}{4}, \quad \frac{\pi}{4}, \quad \frac{5\pi}{4}, \quad \frac{9\pi}{4}, \ldots \quad \text{(in radians).}$$

Only one of these roots, namely 45° or $\frac{\pi}{4}$, is denoted by the function arctan1. This is the value which your calculator will give you. It is called the *principal value.*

The principal value for any inverse trigonometric function is unique and lies within a specified range:

$$-\frac{\pi}{2} < \arctan x < \frac{\pi}{2}$$

$$-\frac{\pi}{2} \leqslant \arcsin x \leqslant \frac{\pi}{2}$$

$$0 \leqslant \arccos x \leqslant \pi.$$

It is possible to deduce all other roots from the principal value and this is shown below.

**tan$\theta$ = *c***

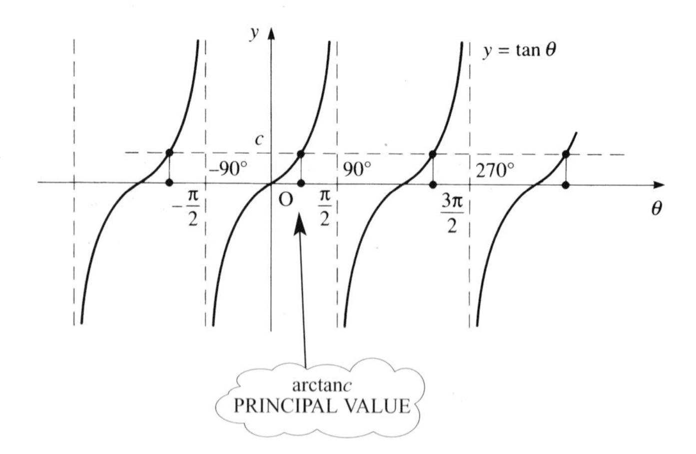


**Figure 2.16**

To solve the equation $\tan\theta = c$, notice how all possible values of $\theta$ occur at intervals of 180° or $\pi$ radians (see figure 2.16). So the general solution is

$$\theta = \arctan c + n\pi \quad n \in \mathbb{Z} \qquad \text{(in radians).}$$

**cos$\theta$ = $c$ for −1 ⩽ $c$ ⩽ 1**

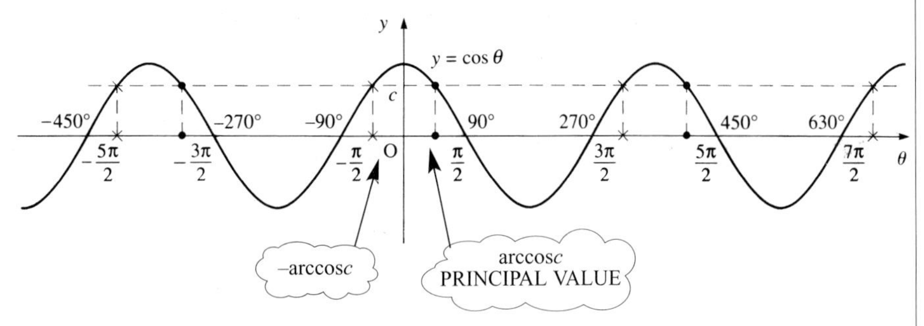


**Figure 2.17**

The cosine graph (see figure 2.17) has the $y$ axis as a line of symmetry. Notice how the values $\pm$ arccos$c$ generate all the other roots at intervals of 360° or $2\pi$. So the general solution is

$$\theta = \pm \operatorname{arccos} c + 2n\pi \qquad n \in \mathbb{Z} \qquad \text{(in radians)}.$$

**sin$\theta$ = $c$ for −1 ⩽ $c$ ⩽ 1**

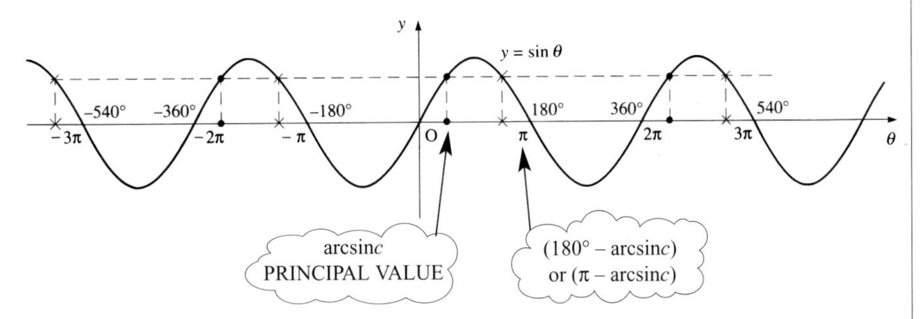


**Figure 2.18**

As in the previous case, there are two roots located symmetrically each side of $\theta = \frac{\pi}{2}$, which generate all the other possible roots (see figure 2.18). This gives rise to the slightly more complicated expressions

$$\theta = \frac{\pi}{2} \pm \left(\frac{\pi}{2} - \operatorname{arcsin} c\right) + 2n\pi$$

or $$\theta = \left(2n + \frac{1}{2}\right)\pi \pm \left(\frac{\pi}{2} - \operatorname{arcsin} c\right) \qquad n \in \mathbb{Z}.$$

You may, however, find it easier to remember these as two separate formulae:

$$\theta = 2n\pi + \operatorname{arcsin} c \qquad \text{or} \qquad \theta = (2n+1)\pi - \operatorname{arcsin} c.$$

**ACTIVITY**

Show that the general solution of the equation $\sin\theta = c$ may also be written

$$\theta = n\pi + (-1)^n \arcsin c.$$

**KEY POINTS**

**1 Compound-angle formulae**

- $\sin(\theta + \phi) = \sin\theta\cos\phi + \cos\theta\sin\phi$
- $\sin(\theta - \phi) = \sin\theta\cos\phi - \cos\theta\sin\phi$
- $\cos(\theta + \phi) = \cos\theta\cos\phi - \sin\theta\sin\phi$
- $\cos(\theta - \phi) = \cos\theta\cos\phi + \sin\theta\sin\phi$
- $\tan(\theta + \phi) = \dfrac{\tan\theta + \tan\phi}{1 - \tan\theta\tan\phi}$
- $\tan(\theta - \phi) = \dfrac{\tan\theta - \tan\phi}{1 + \tan\theta\tan\phi}$

**2 Double-angle and related formulae**

- $\sin 2\theta = 2\sin\theta\cos\theta$
- $\cos 2\theta = \cos^2\theta - \sin^2\theta = 1 - 2\sin^2\theta = 2\cos^2\theta - 1$
- $\tan 2\theta = \dfrac{2\tan\theta}{1 - \tan^2\theta}$
- $\sin^2\theta = \frac{1}{2}(1 - \cos 2\theta)$
- $\cos^2\theta = \frac{1}{2}(1 + \cos 2\theta)$

**3 Factor formulae**

- $\sin\alpha + \sin\beta = 2\sin\left(\dfrac{\alpha + \beta}{2}\right)\cos\left(\dfrac{\alpha - \beta}{2}\right)$
- $\sin\alpha - \sin\beta = 2\cos\left(\dfrac{\alpha + \beta}{2}\right)\sin\left(\dfrac{\alpha - \beta}{2}\right)$
- $\cos\alpha + \cos\beta = 2\cos\left(\dfrac{\alpha + \beta}{2}\right)\cos\left(\dfrac{\alpha - \beta}{2}\right)$
- $\cos\alpha - \cos\beta = -2\sin\left(\dfrac{\alpha + \beta}{2}\right)\sin\left(\dfrac{\alpha - \beta}{2}\right)$

Note the minus sign here

**4 The $r$, $\alpha$ formulae**

- $a\sin\theta + b\cos\theta = r\sin(\theta + \alpha)$
- $a\sin\theta - b\cos\theta = r\sin(\theta + \alpha)$
- $a\cos\theta + b\sin\theta = r\cos(\theta + \alpha)$
- $a\cos\theta - b\sin\theta = r\cos(\theta + \alpha)$

where $r = \sqrt{a^2 + b^2}$

$\cos\alpha = \dfrac{a}{r}$

$\sin\alpha = \dfrac{b}{r}$

**5 The small-angle approximations (for $\theta$ in radians)**

- $\sin\theta \approx \theta$
- $\tan\theta \approx \theta$
- $\cos\theta \approx 1 - \dfrac{\theta^2}{2}$
- $\lim\limits_{\theta \to 0} \dfrac{\theta}{\sin\theta} = \lim\limits_{\theta \to 0} \dfrac{\sin\theta}{\theta} = 1$

# 3 Calculus techniques

**Little by little does the trick.**

*Aesop*

Figure 3.1 depicts a small boat sailing through long swell waves out at sea. How does the angle of the boat to the horizontal vary with the position of the boat along the wave?

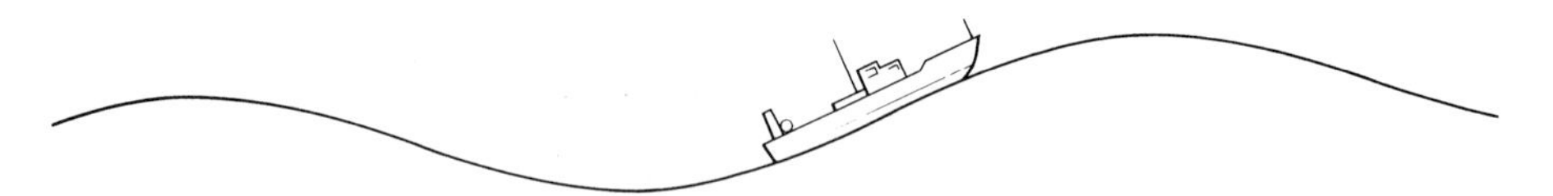

**Figure 3.1**

What calculus techniques would help you to answer this question?

In *Pure Mathematics 1* and *2* you learnt how to differentiate and integrate a variety of functions. These included some compound functions, which were differentiated by using the product rule, quotient rule or chain rule, and integrated by using substitution.

In this chapter, the range of functions which you can differentiate and integrate is extended.

ACTIVITY

**(i)** Figure 3.2 shows the graph of $y = \sin x$, with $x$ measured in radians. You are going to sketch the graph of the gradient function for this graph.

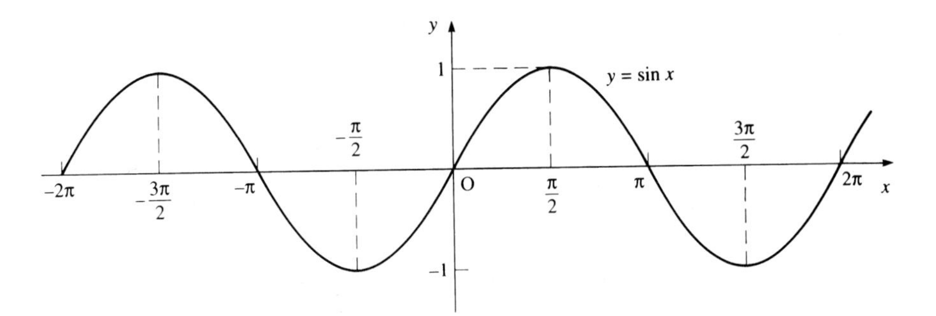


**Figure 3.2**

Draw a horizontal axis for the angles, marked from $-2\pi$ to $2\pi$, and a vertical axis for the gradient, marked from $-1$ to 1.

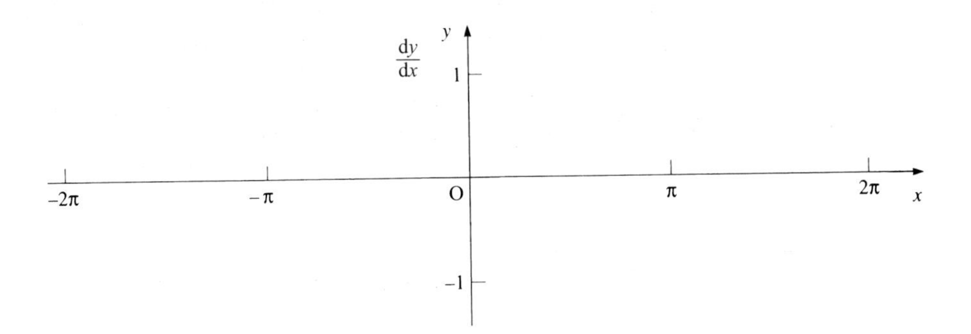


**Figure 3.3**

First, look for the angles for which the gradient of $y = \sin x$ is 0. Mark zeros at these angles on your gradient graph.

Decide which parts of $y = \sin x$ have a positive gradient and which have a negative gradient. This will tell you whether your gradient graph should be above or below the $y$ axis at any point.

Look at the part of the graph near $x = 0$. What do you think the gradient is at this point? (Remember that when $x$ is small, $\sin x \approx x$.) Mark this point on your gradient graph. Also mark on any other points with plus or minus the same gradient.

Now, by considering whether the gradient of $y = \sin x$ is increasing or decreasing at any particular point, sketch in the rest of the gradient graph.

The gradient graph that you have drawn should look like a familiar graph. What graph do you think it is?

**(ii)** Sketch the graph of $y = \cos x$, with $x$ measured in radians, and use it as in part (i) to obtain a sketch of the graph of the gradient function of $y = \cos x$.

## Differentiating *y* = sin*x* from first principles

The activity that you have just done has led you to form an idea of what the derivative of sin$x$ might be. You can prove this by using differentiation from first principles (which you have already met in *Pure Mathematics 1* and *2*).

Figure 3.4 shows part of the graph of $y = \sin x$. The point P is a general point $(x, \sin x)$ on the graph. The point Q is a very small distance further on, so it has $x$ co-ordinate $x + \delta x$, where $\delta x$ is very small, and $y$ co-ordinate $\sin(x + \delta x)$.

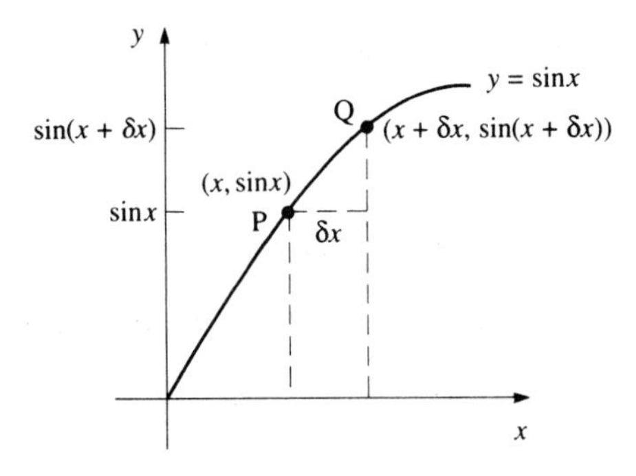


**Figure 3.4**

You can find the gradient at the point P by finding the limit of the gradient of the chord PQ as $\delta x$ approaches 0:

$$\frac{dy}{dx} = \lim_{\delta x \to 0} \frac{\sin(x + \delta x) - \sin x}{\delta x}.$$

$\sin(x + \delta x)$ may be simplified by using the compound-angle formula:

$$\sin(x + \delta x) = \sin x \cos\delta x + \cos x \sin\delta x.$$

As $\delta x$ is small, you can replace $\cos\delta x$ and $\sin\delta x$ by their small-angle approximations:

$$\cos\delta x \approx 1 - \tfrac{1}{2}(\delta x)^2 \qquad \sin\delta x \approx \delta x$$

which leads to

$$\begin{aligned}\sin(x + \delta x) &\approx (\sin x)[1 - \tfrac{1}{2}(\delta x)^2] + (\cos x)\delta x \\ &= \sin x - \tfrac{1}{2}(\sin x)(\delta x)^2 + (\cos x)\delta x.\end{aligned}$$

Substituting this in the expression $\frac{\sin(x+\delta x)-\sin x}{\delta x}$ gives

$$\frac{\sin x - \frac{1}{2}(\sin x)(\delta x)^2 + (\cos x)\delta x - \sin x}{\delta x}$$

$$= \frac{-\frac{1}{2}(\sin x)(\delta x)^2 + (\cos x)\delta x}{\delta x}$$

$$= -\tfrac{1}{2}(\sin x)\delta x + \cos x.$$

In the limit as $\delta x \to 0$, this becomes simply $\cos x$. So

$$\frac{dy}{dx} = \cos x.$$

You have now proved the result which you probably found from the gradient graph sketching activity.

**ACTIVITY**

An alternative way of finding

$$\lim_{\delta x \to 0} \frac{\sin(x+\delta x)-\sin x}{\delta x}$$

involves using the factor formula

$$\sin\theta - \sin\phi = 2\sin\tfrac{1}{2}(\theta-\phi)\cos\tfrac{1}{2}(\theta+\phi).$$

Show that this leads to the same result for the derivative of $\sin x$.

Remember that the small-angle approximations are only valid when the angle is measured in radians. This means that you can only use the standard results for the derivatives of trigonometric functions if you are working in radians.

**ACTIVITY**

**(i)** Use differentiation from first principles to show that

$$\frac{d}{dx}(\cos x) = -\sin x.$$

**(ii)** By writing $\tan x = \frac{\sin x}{\cos x}$, use the quotient rule to show that

$$\frac{d}{dx}(\tan x) = \sec^2 x.$$

## Summary of results

You now know how to differentiate the functions $\sin x$, $\cos x$ and $\tan x$.

$$\frac{d}{dx}(\sin x) = \cos x$$

$$\frac{d}{dx}(\cos x) = -\sin x$$

$$\frac{d}{dx}(\tan x) = \sec^2 x$$

where $x$ is measured in radians.

You can use these results to differentiate a variety of functions involving trigonometric functions, by using the product rule, quotient rule or chain rule as appropriate, as in the following examples.

**EXAMPLE 3.1**

Differentiate $y = \cos 2x$.

**SOLUTION**

As $\cos 2x$ is a function of a function, you may use the chain rule.

$$\text{Let } u = 2x \quad \Rightarrow \quad \frac{du}{dx} = 2$$

$$y = \cos u \Rightarrow \quad \frac{dy}{du} = -\sin u$$

$$\frac{dy}{dx} = \frac{dy}{du} \times \frac{du}{dx}$$

$$= -\sin u \times 2$$

$$= -2\sin 2x.$$

With practice it should be possible to do this in your head, without needing to write down the substitution.

This result may be generalised:

$$y = \cos kx \Rightarrow \frac{dy}{dx} = -k\sin kx.$$

$$\text{Similarly } y = \sin kx \Rightarrow \frac{dy}{dx} = k\cos kx.$$

**EXAMPLE 3.2**

Differentiate $y = x^2\sin x$.

**SOLUTION**

$x^2\sin x$ is of the form $uv$, so the product rule can be used with $u = x^2$ and $v = \sin x$.

$$\frac{du}{dx} = 2x \qquad \frac{dv}{dx} = \cos x$$

The product rule

$$\frac{dy}{dx} = v\frac{du}{dx} + u\frac{dv}{dx}$$

$$\Rightarrow \quad \frac{dy}{dx} = 2x\sin x + x^2\cos x.$$

**EXAMPLE 3.3**

Differentiate $y = e^{\tan x}$.

**SOLUTION**

$e^{\tan x}$ is a function of a function, so the chain rule may be used.

Let $u = \tan x \quad \Rightarrow \quad \frac{du}{dx} = \sec^2 x$

$y = e^u \quad \Rightarrow \quad \frac{dy}{du} = e^u.$

Using the chain rule

$$\frac{dy}{dx} = \frac{dy}{du} \times \frac{du}{dx}$$

$$= e^u \sec^2 x$$

$$= \sec^2 x \, e^{\tan x}.$$

**EXAMPLE 3.4**

Differentiate $y = \dfrac{1 + \sin x}{\cos x}$.

**SOLUTION**

$\dfrac{1 + \sin x}{\cos x}$ is of the form $\dfrac{u}{v}$ so the quotient rule can be used, with

$$u = 1 + \sin x \quad \text{and} \quad v = \cos x$$

$$\Rightarrow \quad \frac{du}{dx} = \cos x \qquad \frac{dv}{dx} = -\sin x.$$

The quotient rule is

$$\frac{dy}{dx} = \frac{v\frac{du}{dx} - u\frac{dv}{dx}}{v^2}.$$

Substituting for $u$ and $v$ and their derivatives gives

$$\frac{dy}{dx} = \frac{(\cos x)(\cos x) - (1 + \sin x)(-\sin x)}{(\cos x)^2}$$

$$= \frac{\cos^2 x + \sin x + \sin^2 x}{\cos^2 x}$$

$$= \frac{1 + \sin x}{\cos^2 x} \qquad \text{(using } \sin^2 x + \cos^2 x = 1\text{).}$$

## EXERCISE 3A

**1** Differentiate each of the following functions.

**(i)** $2\cos x + \sin x$ **(ii)** $\tan x + 5$ **(iii)** $\sin x - \cos x$

**2** Use the product rule to differentiate each of the following functions.

**(i)** $x\tan x$ **(ii)** $\sin x\cos x$ **(iii)** $e^x\sin x$

**3** Use the quotient rule to differentiate each of the following functions:

**(i)** $\dfrac{\sin x}{x}$ **(ii)** $\dfrac{e^x}{\cos x}$ **(iii)** $\dfrac{x + \cos x}{\sin x}$

**4** Use the chain rule to differentiate each of the following functions.

**(i)** $\tan(x^2 + 1)$ **(ii)** $\cos^2 x$ **(iii)** $\ln(\sin x)$

**5** Use an appropriate method to differentiate each of the following functions.

**(i)** $\sqrt{\cos x}$ **(ii)** $e^x\tan x$ **(iii)** $\sin 4x^2$

**(iv)** $e^{\cos 2x}$ **(v)** $\dfrac{\sin x}{1 + \cos x}$ **(vi)** $\ln(\tan x)$

**6** **(i)** By writing $\sec x$ as $\dfrac{1}{\cos x}$, differentiate $\sec x$.

**(ii)** By writing $\operatorname{cosec} x$ as $\dfrac{1}{\sin x}$, differentiate $\operatorname{cosec} x$.

**(iii)** By writing $\cot x$ as $\dfrac{\cos x}{\sin x}$, differentiate $\cot x$.

**7** **(i)** Differentiate $y = x\cos x$.

**(ii)** Find the gradient of the curve $y = x\cos x$ at the point where $x = \pi$.

**(iii)** Find the equation of the tangent to the curve $y = x\cos x$ at the point where $x = \pi$.

**(iv)** Find the equation of the normal to the curve $y = x\cos x$ at the point where $x = \pi$.

**8** The function $y = \sin^3 x$ has five stationary points in $-\pi \leqslant x \leqslant \pi$.

**(i)** Find $\dfrac{dy}{dx}$ for this function.

**(ii)** Find the co-ordinates of the five stationary points.

**(iii)** Determine whether each of the five points is a maximum, minimum or point of inflection.

**(iv)** Use this information to sketch the graph of $y = \sin^3 x$ for values of $x$ in $-\pi \leqslant x \leqslant \pi$.

**9** For the curve $y = x + \sin 2x$:

**(i)** find $\frac{dy}{dx}$;

**(ii)** find the co-ordinates of the stationary points in $0 \leqslant x \leqslant 2\pi$, and determine their nature;

**(iii)** sketch the curve for $0 \leqslant x \leqslant 2\pi$.

**10** If $y = e^x \cos 3x$, find $\frac{dy}{dx}$ and $\frac{d^2y}{dx^2}$ and hence show that

$$\frac{d^2y}{dx^2} - 2\frac{dy}{dx} + 10y = 0.$$

[MEI]

**11** Consider the function $y = e^{-x}\sin x$, where $-\pi \leqslant x \leqslant \pi$.

**(i)** Find $\frac{dy}{dx}$.

**(ii)** Show that, at stationary points, $\tan x = 1$.

**(iii)** Determine the co-ordinates of the stationary points, correct to 2 significant figures.

**(iv)** Explain how you could determine whether your stationary points are maxima or minima. You are not required to do any calculations.

[MEI]

**12 (i) (a)** Show that $(\cos x + \sin x)^2 = 1 + \sin 2x$, for all $x$.

**(b)** Hence, or otherwise, find the derivative of $(\cos x + \sin x)^2$.

**(ii) (a)** By expanding $(\cos^2 x + \sin^2 x)^2$, find and simplify an expression for $\cos^4 x + \sin^4 x$ involving $\sin 2x$.

**(b)** Hence, or otherwise, show that the derivative of $\cos^4 x + \sin^4 x$ is $-\sin 4x$.

[MEI]

**13** You are given that $y = e^{-x}\sin 2x$.

**(i)** Find $\frac{dy}{dx}$.

**(ii)** Show that, at stationary points, $\tan 2x = 2$.

**(iii)** Find the solutions in radians of the equation $\tan 2x = 2$ which lie in the range $0 \leqslant x \leqslant \pi$ correct to 2 decimal places.

**(iv)** Show that $\frac{dy}{dx}$ can be written as $re^{-x}\cos(2x + \alpha)$ where $r$ and $\alpha$ are to be determined.

[MEI]

**14** You are given that

$$y = e^{-2x}\tan x.$$

Find $\frac{dy}{dx}$, and show that at all stationary points $\sin 2x = 1$.

[MEI]

# Differentiating functions defined implicitly

All the functions you have differentiated so far have been of the form $y = \mathrm{f}(x)$. However, many functions cannot be arranged in this way at all, for example $x^3 + y^3 = xy$, and others can look clumsy when you try to make $y$ the subject.

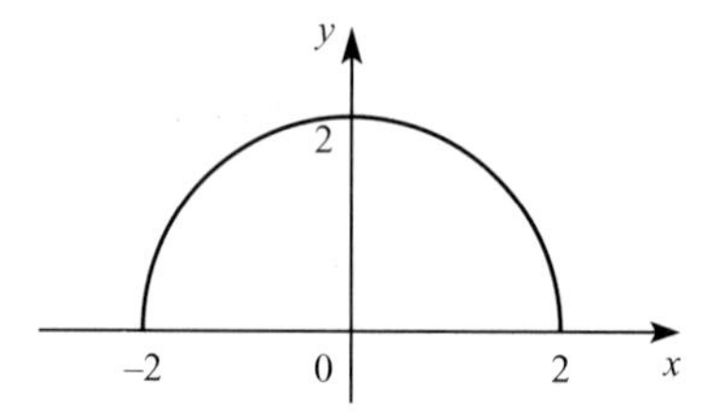


**Figure 3.5**

An example of this is the semi-circle $x^2 + y^2 = 4$, $y \geqslant 0$, illustrated in figure 3.5.

The curve is much more easily recognised in this form than in the equivalent $y = \sqrt{4 - x^2}$.

When a function is specified by an equation connecting $x$ and $y$ which does not have $y$ as the subject it is called an *implicit function*.

---

Although restrictions on $x$ or $y$ are often necessary to make the function unambiguous, we frequently assume such restrictions but do not mention them.

---

The chain rule $\frac{\mathrm{d}y}{\mathrm{d}x} = \frac{\mathrm{d}y}{\mathrm{d}u} \times \frac{\mathrm{d}u}{\mathrm{d}x}$ and the product rule $\frac{\mathrm{d}}{\mathrm{d}x}(uv) = u\frac{\mathrm{d}v}{\mathrm{d}x} + v\frac{\mathrm{d}u}{\mathrm{d}x}$ are used extensively to help in the differentiation of implicit functions.

**EXAMPLE 3.5**

Differentiate each of the following with respect to $x$:

**(i)** $y^2$ **(ii)** $xy$ **(iii)** $3x^2y^3$ **(iv)** $\sin y$.

**SOLUTION**

**(i)** $$\frac{\mathrm{d}}{\mathrm{d}x}(y^2) = \frac{\mathrm{d}}{\mathrm{d}y}(y^2) \times \frac{\mathrm{d}y}{\mathrm{d}x} \text{ (chain rule)}$$
$$= 2y\frac{\mathrm{d}y}{\mathrm{d}x}$$

**(ii)** $$\frac{\mathrm{d}}{\mathrm{d}x}(xy) = x\frac{\mathrm{d}y}{\mathrm{d}x} + y \quad \text{(product rule)}$$

**(iii)** $\frac{d}{dx}(3x^2y^3) = 3\left(x^2\frac{d}{dx}(y^3) + y^3\frac{d}{dx}(x^2)\right)$ (product rule)

$= 3\left(x^2 \times 3y^2\frac{dy}{dx} + y^3 \times 2x\right)$ (chain rule)

$= 3xy^2\left(3x\frac{dy}{dx} + 2y\right)$

**(iv)** $\frac{d}{dx}(\sin y) = \frac{d}{dy}(\sin y) \times \frac{dy}{dx}$ (chain rule)

$= (\cos y)\frac{dy}{dx}$

**EXAMPLE 3.6**

The equation of a curve is given by $y^3 + xy = 2$.

**(i)** Find an expression for $\frac{dy}{dx}$ in terms of $x$ and $y$.

**(ii)** Hence find the gradient of the curve at (1, 1) and the equation of the tangent to the curve at that point.

**SOLUTION**

**(i)** $y^3 + xy = 2$

$\Rightarrow \quad 3y^2\frac{dy}{dx} + (x\frac{dy}{dx} + y) = 0$

$\Rightarrow \quad (3y^2 + x)\frac{dy}{dx} = -y$

$\Rightarrow \quad \frac{dy}{dx} = \frac{-y}{(3y^2 + x)}.$

**(ii)** At (1, 1), $\frac{dy}{dx} = -\frac{1}{4}$

$\Rightarrow$ using $y - y_1 = m(x - x_1)$ the equation of the tangent is $(y - 1) = -\frac{1}{4}(x - 1)$

$\Rightarrow \quad x + 4y - 5 = 0.$

---

**?** Figure 3.6 shows the graph of this function.

**(i)** How can you use your graphics calculator to sketch this? (**Hint:** What effect does interchanging $x$ and $y$ have on a graph?)

**(ii)** Why is this not a function?

---

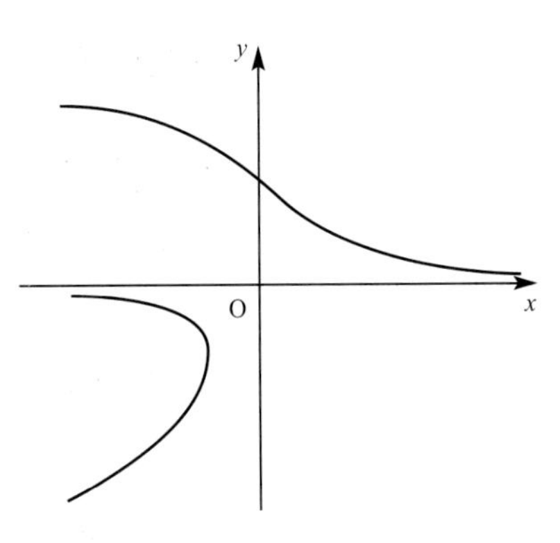


**Figure 3.6**

## Stationary points

As before these occur where $\frac{dy}{dx} = 0$.

Putting $\frac{dy}{dx} = 0$ will not usually give values of $x$ directly, but will give a relationship between $x$ and $y$. This needs to be solved simultaneously with the equation of the curve to find the co-ordinates.

**EXAMPLE 3.7**

**(i)** Differentiate $x^3 + y^3 = 3xy$ with respect to $x$.

**(ii)** Hence find the co-ordinates of any stationary points.

**SOLUTION**

**(i)** $$\frac{d}{dx}(x^3) + \frac{d}{dx}(y^3) = \frac{d}{dx}(3xy)$$

$$\Rightarrow \quad 3x^2 + 3y^2\frac{dy}{dx} = 3\left[x\frac{dy}{dx} + y\right].$$

**(ii)** At stationary points, $\frac{dy}{dx} = 0$

$$\Rightarrow \quad 3x^2 = 3y$$

Notice how it is not necessary to find an expression for $\frac{dy}{dx}$ unless you are told to

$$\Rightarrow \quad x^2 = y$$

To find the co-ordinates of the stationary points, solve

$$\left.\begin{array}{l} x^2 = y \\ x^3 + y^3 = 3xy \end{array}\right\} \text{ simultaneously}$$

Substituting for $y$ we obtain

$$x^3 + (x^2)^3 = 3x(x^2)$$

$$\Rightarrow \quad x^3 + x^6 = 3x^3$$

$$\Rightarrow \quad x^6 = 2x^3$$

$$\Rightarrow \quad x^3(x^3 - 2) = 0$$

$$\Rightarrow \quad x = 0 \quad \text{or} \quad x = \sqrt[3]{2}$$

$y = x^2 \quad \Rightarrow$ stationary points are $(0, 0)$ and $(\sqrt[3]{2}, \sqrt[3]{4})$.

The stationary points are A and B in figure 3.7.

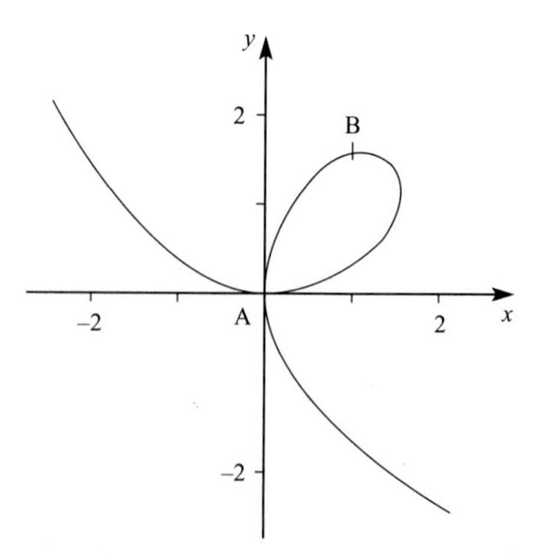


**Figure 3.7**

## Types of stationary points

As with explicit functions, the type of stationary point can usually be determined by considering the sign of the second derivative $\frac{d^2y}{dx^2}$ at the stationary point.

**EXAMPLE 3.8**

The curve with equation $\sin x + \sin y = 1$ for $0 \leqslant x \leqslant \pi$, $0 \leqslant y \leqslant \pi$ is shown in figure 3.8.

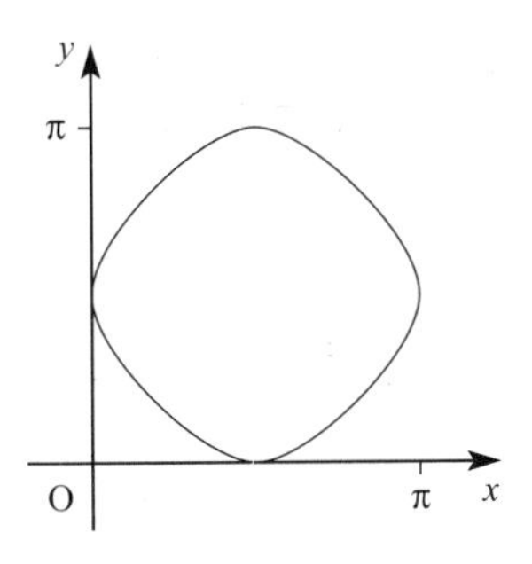


**Figure 3.8**

**(i)** Differentiate the equation of the curve with respect to $x$ and hence find the co-ordinates of any stationary points.

**(ii)** Differentiate the equation again with respect to $x$ to determine the types of stationary points.

**SOLUTION**

**(i)** $\sin x + \sin y = 1$

$$\Rightarrow \quad \cos x + (\cos y)\frac{dy}{dx} = 0 \qquad ①$$

$$\Rightarrow \quad \frac{dy}{dx} = -\frac{\cos x}{\cos y}.$$

At any stationary points $\frac{dy}{dx} = 0 \quad \Rightarrow \quad \cos x = 0$

$\Rightarrow \quad x = \frac{\pi}{2}$ (only solution in range)

Substitute in $\sin x + \sin y = 1$.

When $x = \frac{\pi}{2}$, $\sin x = 1 \quad \Rightarrow \quad \sin y = 0$

$\Rightarrow \quad y = 0$ or $y = \pi$

$\Rightarrow$ turning points at $(\frac{\pi}{2}, 0)$ and $(\frac{\pi}{2}, \pi)$.

**(ii)** Differentiating equation ① again with respect to $x$:

$$\cos x + (\cos y)\frac{dy}{dx} = 0$$

$$\Rightarrow \quad -\sin x + \left[(\cos y)\frac{d^2y}{dx^2} + \frac{dy}{dx}\left((-\sin y)\frac{dy}{dx}\right)\right] = 0$$

At $(\frac{\pi}{2}, 0)$, $\frac{dy}{dx} = 0$

$$\Rightarrow \quad -\sin\tfrac{\pi}{2} + (\cos 0)\frac{d^2y}{dx^2} = 0$$

$$\Rightarrow \quad \frac{d^2y}{dx^2} = 1 \;\Rightarrow\; \text{minimum turning point at } (\tfrac{\pi}{2}, 0).$$

At $(\frac{\pi}{2}, \pi)$, $\frac{dy}{dx} = 0$

$$\Rightarrow \quad -\sin\tfrac{\pi}{2} + (\cos\pi)\frac{d^2y}{dx^2} = 0$$

$$\Rightarrow \quad -1 - \frac{d^2y}{dx^2} = 0$$

$$\Rightarrow \quad \frac{d^2y}{dx^2} = -1 \;\Rightarrow\; \text{maximum turning point at } (\tfrac{\pi}{2}, \pi).$$

These points are confirmed by considering the sketch in figure 3.8.

**EXERCISE 3B**

**1** Differentiate each of the following with respect to $x$.

**(i)** $y^4$ **(ii)** $x^2 + y^3 - 5$ **(iii)** $xy + x + y$
**(iv)** $\cos y$ **(v)** $e^{(y+2)}$ **(vi)** $xy^3$
**(vii)** $2x^2y^5$ **(viii)** $x + \ln y - 3$ **(ix)** $xe^y - \cos y$
**(x)** $x^2 \ln y$ **(xi)** $xe^{\sin y}$ **(xii)** $x\tan y - y\tan x$

**2** Find the gradient of the curve $xy^3 = 5\ln y$ at the point $(0, 1)$.

**3** Find the gradient of the curve $e^{\sin x} + e^{\cos y} = e + 1$ at the point $(\frac{\pi}{2}, \frac{\pi}{2})$.

**4** **(i)** Find the gradient of the curve $x^2 + 3xy + y^2 = x + 3y$ at the point $(2, 0)$.
**(ii)** Hence find the equation of the tangent to the curve at this point.

**5** Find the co-ordinates of all the stationary points on the curve $x^2 + y^2 + xy = 3$.

**6** **(i)** Show that the graph of $xy + 48 = x^2 + y^2$ has stationary points at $(4, 8)$ and $(-4, -8)$.
**(ii)** By differentiating with respect to $x$ a second time determine the nature of these stationary points.

**7** A curve has equation $(x-6)(y+4)=2$.

**(i)** Find an expression for $\frac{dy}{dx}$ in terms of $x$ and $y$.

**(ii)** Find the equation of the normal to the curve at the point $(7, -2)$.

**(iii)** Find the co-ordinates of the point where the normal meets the curve again.

**(iv)** By rewriting the equation in the form $y - a = \frac{b}{x-c}$ identify any asymptotes and sketch the curve.

**8** A curve has equation $y = x^x$ for $x > 0$.

**(i)** Take logarithms to base e of both sides of the equation.

**(ii)** Differentiate the resulting equation with respect to $x$.

**(iii)** Find the co-ordinates of the stationary point, giving your answer to 3 decimal places.

**(iv)** Sketch the curve for $x > 0$.

## Integrating sin$x$ and cos$x$

Since $\frac{d}{dx}(\sin x) = \cos x$

it follows that $\int \cos x \, dx = \sin x + c$

Similarly, since $\frac{d}{dx}(\cos x) = -\sin x$

it also follows that $\frac{d}{dx}(-\cos x) = \sin x$

and therefore $\int \sin x \, dx = -\cos x + c$

With this knowledge, you can now integrate not only the functions sin$x$ and cos$x$, but also many other functions by using substitution.

**EXAMPLE 3.9**

Find $\int \sin 7x \, dx$.

**SOLUTION**

Make the substitution $u = 7x$. Then differentiate:

$$\frac{du}{dx} = 7$$

$$\Rightarrow \quad dx = \tfrac{1}{7} du.$$

$$\int \sin 7x \, dx = \int \tfrac{1}{7} \sin u \, du$$

$$= -\tfrac{1}{7} \cos u + c$$

$$= -\tfrac{1}{7} \cos 7x + c.$$

*Note*

You would not usually use a substitution for an integral like this but would quote the general result that

$$\int \sin kx \, dx = -\frac{1}{k}\cos kx + c.$$

Similarly $\int \cos kx \, dx = \frac{1}{k}\sin kx + c.$

**EXAMPLE 3.10**

Find $\int 2x\cos(x^2 + 1)\, dx$.

**SOLUTION**

Make the substitution $u = x^2 + 1$. Then differentiate:

$$\frac{du}{dx} = 2x$$

$$\Rightarrow \quad 2x\,dx = du.$$

$$\int 2x\cos(x^2 + 1)\,dx = \int \cos u\,du$$
$$= \sin u + c$$
$$= \sin(x^2 + 1) + c.$$

Notice that the last example involves two functions of $x$ multiplied together, namely $2x$ and $\cos(x^2 + 1)$. These two functions are related by the fact that $2x$ is the derivative of $x^2 + 1$. Because of this relationship, the substitution $u = x^2 + 1$ may be used to perform the integration. You have already met integrals of this type in *Pure Mathematics 2*, and you can now apply this to some integrals involving trigonometric functions.

**EXAMPLE 3.11**

Find $\int_0^{\frac{\pi}{2}} \cos x \sin^2 x\, dx$.

(Remember that $\sin^2 x$ means the same as $(\sin x)^2$.)

**SOLUTION**

This integral is the product of two functions, $\cos x$ and $(\sin x)^2$.

Now $(\sin x)^2$ is a function of $\sin x$, and $\cos x$ is the derivative of $\sin x$, so you should use the substitution $u = \sin x$.

Differentiating:

$$\frac{du}{dx} = \cos x \quad \Rightarrow \quad du = \cos x\,dx.$$

The limits of integration need to be changed as well:

$$x = \tfrac{\pi}{2} \quad \Rightarrow \quad u = 1$$

$$x = 0 \quad \Rightarrow \quad u = 0$$

Therefore $\displaystyle\int_0^{\frac{\pi}{2}} \cos x \sin^2 x \, dx = \int_0^1 u^2 du$

$$= \left[\frac{u^3}{3}\right]_0^1$$

$$= \tfrac{1}{3}.$$

***Note***

You may find that as you gain practice in this type of integration you become able to work out the integral without writing down the substitution. However, if you are unsure, it is best to write down the whole process.

Another special case of integration by substitution which you met in *Pure Mathematics 2* is that used for integrals of the form

$$\int \frac{f'(x)}{f(x)} \, dx.$$

This integral is equal to $\ln|f(x)| + c$, as is shown in the following example.

**EXAMPLE 3.12**

Find $\displaystyle\int \frac{\cos x}{1 + \sin x} \, dx$.

**SOLUTION**

Let $u = 1 + \sin x$.

$$\frac{du}{dx} = \cos x \quad \Rightarrow \quad du = \cos x \, dx$$

$$\int \frac{\cos x}{1 + \sin x} dx \qquad = \int \frac{1}{u} \, du$$

$$= \ln|u| + c$$

$$= \ln|1 + \sin x| + c$$

Again, you may feel able to do this without writing down the substitution. It is an example of the rule 'If you obtain the top line when you differentiate the bottom line, the integral is the natural logarithm of the modulus of the bottom line.'

## Using trigonometric identities in integration

Sometimes, when it is not immediately obvious how to integrate a function involving trigonometric functions, it may help to rewrite the function using one of the trigonometric identities.

**EXAMPLE 3.13**

Find $\int \sin^2 x \, dx$.

**SOLUTION**

A substitution cannot be used in this case. However, in Chapter 2 you learnt the identity

$$\cos 2x = 1 - 2\sin^2 x.$$

(Remember that this is just one of the three expressions for $\cos 2x$.)

This identity may be rewritten as

$$\sin^2 x = \tfrac{1}{2}(1 - \cos 2x).$$

By putting $\sin^2 x$ in this form, you will be able to perform the integration:

$$\begin{aligned}\int \sin^2 x \, dx &= \tfrac{1}{2}\int (1 - \cos 2x) \, dx \\ &= \tfrac{1}{2}(x - \tfrac{1}{2}\sin 2x) + c \\ &= \tfrac{1}{2}x - \tfrac{1}{4}\sin 2x + c.\end{aligned}$$

You can integrate $\cos^2 x$ in the same way, by using $\cos^2 x = \tfrac{1}{2}(\cos 2x + 1)$. Other even powers of $\sin x$ or $\cos x$ can also be integrated in a similar way, but you have to use the identity twice or more.

**EXAMPLE 3.14**

Find $\int \cos^4 x \, dx$.

**SOLUTION**

First express $\cos^4 x$ as $(\cos^2 x)^2$:

$$\begin{aligned}\cos^4 x &= [\tfrac{1}{2}(\cos 2x + 1)]^2 \\ &= \tfrac{1}{4}(\cos^2 2x + 2\cos 2x + 1).\end{aligned}$$

Next, apply the same identity to $\cos^2 2x$:

$$\cos^2 2x = \tfrac{1}{2}(\cos 4x + 1).$$

Hence

$$\begin{aligned}\cos^4 x &= \tfrac{1}{4}(\tfrac{1}{2}\cos 4x + \tfrac{1}{2} + 2\cos 2x + 1) \\ &= \tfrac{1}{4}(\tfrac{1}{2}\cos 4x + 2\cos 2x + \tfrac{3}{2}) \\ &= \tfrac{1}{8}\cos 4x + \tfrac{1}{2}\cos 2x + \tfrac{3}{8}.\end{aligned}$$

This can now be integrated:

$$\int \cos^4 x \,dx = \int (\tfrac{1}{8}\cos 4x + \tfrac{1}{2}\cos 2x + \tfrac{3}{8}) \,dx$$
$$= \tfrac{1}{32}\sin 4x + \tfrac{1}{4}\sin 2x + \tfrac{3}{8}x + c.$$

For odd powers of $\sin x$ or $\cos x$, a different technique is used, as in the next example.

**EXAMPLE 3.15**

Find $\int \cos^3 x \,dx$.



**SOLUTION**

First write $\cos^3 x = \cos x \cos^2 x$.

Now remember that

$$\cos^2 x + \sin^2 x = 1 \quad \Rightarrow \quad \cos^2 x = 1 - \sin^2 x.$$

This gives

$$\cos^3 x = \cos x(1 - \sin^2 x) = \cos x - \cos x \sin^2 x.$$

The first part of this expression, $\cos x$, is easily integrated to give $\sin x$.

The second part is more complicated, but it can be seen that it is of a type we have met already, as it is a product of two functions, one of which is a function of $\sin x$ and the other of which is the derivative of $\sin x$. This can be integrated either by making the substitution $u = \sin x$ or simply in your head (by inspection). So

$$\int \cos^3 x \,dx = \int (\cos x - \cos x \sin^2 x) \,dx$$
$$= \sin x - \tfrac{1}{3}\sin^3 x + c.$$

Any odd power of $\sin x$ or $\cos x$ can be integrated in this way, but again it may be necessary to use the identity more than once. For example:

$$\sin^5 x = \sin x(\sin^2 x)(\sin^2 x) = \sin x(1 - \cos^2 x)^2$$
$$= \sin x(1 - 2\cos^2 x + \cos^4 x)$$
$$= \sin x - 2\sin x \cos^2 x + \sin x \cos^4 x.$$

This can now be integrated.

**EXERCISE 3C**

**1** Integrate the following functions with respect to $x$.

**(i)** $\sin x - 2\cos x$ **(ii)** $3\cos x + 2\sin x$ **(iii)** $5\sin x + 4\cos x$

**2** Integrate the following functions by using the substitution given, or otherwise.

**(i)** $\cos 3x$ $u = 3x$

**(ii)** $\sin(1 - x)$ $u = 1 - x$

**(iii)** $\sin x \cos^3 x$ $u = \cos x$

**(iv)** $\dfrac{\sin x}{2 - \cos x}$ $u = 2 - \cos x$

**(v)** $\tan x$ $u = \cos x$ $\left(\text{write } \tan x \text{ as } \dfrac{\sin x}{\cos x}\right)$

**(vi)** $\sin 2x(1 + \cos 2x)^2$ $u = 1 + \cos 2x$

**3** Use a suitable substitution to integrate the following functions.

**(i)** $2x\sin(x^2)$ **(ii)** $\cos x e^{\sin x}$

**(iii)** $\sec^2 x \tan x$ **(iv)** $\dfrac{\cos x}{\sin^2 x}$

**4** Evaluate the following definite integrals by using suitable substitutions.

**(i)** $\displaystyle\int_0^{\frac{\pi}{2}} \cos(2x - \tfrac{\pi}{2})\,dx$ **(ii)** $\displaystyle\int_0^{\frac{\pi}{4}} \cos x \sin^3 x\,dx$

**(iii)** $\displaystyle\int_0^{\sqrt{\pi}} x\sin(x^2)\,dx$ **(iv)** $\displaystyle\int_0^{\frac{\pi}{4}} \sec^2 x\, e^{\tan x}\,dx$

**(v)** $\displaystyle\int_0^{\frac{\pi}{4}} \frac{\sec^2 x}{1 + \tan x}\,dx$

**5 (i)** Use a graphics calculator or computer to sketch the graph of the function $y = \sin x(\cos x - 1)^2$ for $0 \leqslant x \leqslant 4\pi$.

**(ii)** Use definite integration to find the area between the positive part of one cycle of the curve and the $x$ axis.

**6** Use a suitable trigonometric identity to perform the following integrations.

**(i)** $\displaystyle\int \cos^2 x\,dx$ **(ii)** $\displaystyle\int \sin^3 x\,dx$

**(iii)** $\displaystyle\int \sin^4 x\,dx$ **(iv)** $\displaystyle\int \cos^5 x\,dx$

## Integration by parts

There are still many integrations which you cannot yet do. In fact, many functions cannot be integrated at all, although virtually all functions can be differentiated. However, some functions can be integrated by techniques which you have not yet met. Some of these will be developed in this chapter, starting with integration by parts.

**EXAMPLE 3.16**

Find $\displaystyle\int x\cos x\,dx$.

**SOLUTION**

The function to be integrated is clearly a product of two simpler functions, $x$ and $\cos x$, so your first thought may be to look for a substitution to enable you to perform the integration. However, there are some functions which are products but which cannot be integrated by substitution. This is one of them. You need a new technique to integrate such functions.

Take the function $x\sin x$ and differentiate it, using the product rule:

$$\frac{d}{dx}(x\sin x) = x\cos x + \sin x.$$

Now integrate both sides. This has the effect of 'undoing' the differentiation, so

$$x\sin x = \int x\cos x \, dx + \int \sin x \, dx.$$

Rearranging this gives

$$\int x\cos x \, dx = x\sin x - \int \sin x \, dx$$

$$= x\sin x - (-\cos x) + c$$

$$= x\sin x + \cos x + c.$$

This has enabled you to find the integral of $x\cos x$.

The work in this example can be generalised into the method of integration by parts. Before coming on to that, do the following activity.

**ACTIVITY**

For each of the following:

**(i)** differentiate the given function $f(x)$ using the product rule;

**(ii)** rearrange your expression to find an expression for the given integral $I$;

**(iii)** use this expression to find the given integral.

**(a)** $f(x) = x\cos x \qquad I = \int x\sin x \, dx$

**(b)** $f(x) = xe^{2x} \qquad I = \int 2x\, e^{2x} \, dx$

---

? The above work has enabled you to work out some integrals which you could not previously have done, but you needed to be given the functions to be differentiated first. Effectively you were given the answers.

Look at the expressions you found in part (ii) of the above activity. Can you see any way of working out these expressions without starting by differentiating a given product?

---

## The general result for integration by parts

You will now generalise the method just investigated.

Look back at Example 3.16. Use $u$ to stand for the function $x$, and $v$ to stand for the function $\sin x$.

Using the product rule to differentiate the function $uv$:

$$\frac{d}{dx}(uv) = v\frac{du}{dx} + u\frac{dv}{dx}$$

Integrating gives

$$uv = \int v\frac{\mathrm{d}u}{\mathrm{d}x}\,\mathrm{d}x + \int u\frac{\mathrm{d}v}{\mathrm{d}x}\,\mathrm{d}x$$

Rearranging gives

$$\int u\frac{\mathrm{d}v}{\mathrm{d}x}\,\mathrm{d}x = uv - \int v\frac{\mathrm{d}u}{\mathrm{d}x}\,\mathrm{d}x$$

This is the expression you use when you need to integrate by parts.

In order to use this, you have to split the function you want to integrate into two simpler functions. In Example 3.16 you would split $x\cos x$ into the two functions $x$ and $\cos x$. One of these functions will be called $u$, and the other $\frac{\mathrm{d}v}{\mathrm{d}x}$, to fit the left-hand side of the expression. You will need to decide which will be which. Two considerations will help you:

- As you want to use $\frac{\mathrm{d}u}{\mathrm{d}x}$ on the right-hand side of the expression, $u$ should be a function which becomes a simpler function after differentiation. So in this case, $u$ will be the function $x$.
- As you need $v$ to work out the right-hand side of the expression, it must be possible to integrate the function $\frac{\mathrm{d}v}{\mathrm{d}x}$ to obtain $v$. In this case, $\frac{\mathrm{d}v}{\mathrm{d}x}$ will be the will be the function $\cos x$.

So now you can find $\int x\cos x\,\mathrm{d}x$.

Put $\quad u = x \quad \Rightarrow \quad \frac{\mathrm{d}u}{\mathrm{d}x} = 1$

and $\quad \frac{\mathrm{d}v}{\mathrm{d}x} = \cos x \quad \Rightarrow \quad v = \sin x$

Substituting in

$$\int u\frac{\mathrm{d}v}{\mathrm{d}x}\,\mathrm{d}x = uv - \int v\frac{\mathrm{d}v}{\mathrm{d}x}\,\mathrm{d}x$$

gives

$$\begin{aligned}\int x\cos x\,\mathrm{d}x &= x\sin x - \int 1 \times \sin x\,\mathrm{d}x \\ &= x\sin x - (-\cos x) + \mathrm{c} \\ &= x\sin x + \cos x + c.\end{aligned}$$

**EXAMPLE 3.17**

Find $\int 2xe^x \, dx$.

**SOLUTION**

First split $2xe^x$ into the two simpler functions, $2x$ and $e^x$. Both can be integrated easily, but as $2x$ becomes a simpler function after differentiation and $e^x$ does not, take $u$ to be $2x$.

$$u = 2x \quad \Rightarrow \quad \frac{du}{dx} = 2$$

$$\frac{dv}{dx} = e^x \quad \Rightarrow \quad v = e^x.$$

Substituting in

$$\int u\frac{dv}{dx}\,dx = uv - \int v\frac{du}{dx}\,dx$$

gives

$$\int 2xe^x\,dx = 2xe^x - \int 2e^x\,dx$$
$$= 2xe^x - 2e^x + c.$$

Sometimes it is necessary to use integration by parts twice or more to complete the integration successfully.

**EXAMPLE 3.18**

Find $\int x^2 \sin x \, dx$.

**SOLUTION**

First split $x^2\sin x$ into two: $x^2$ and $\sin x$. As $x^2$ becomes a simpler function after differentiation, take $u$ to be $x^2$.

$$u = x^2 \quad \Rightarrow \quad \frac{du}{dx} = 2x$$

$$\frac{dv}{dx} = \sin x \quad \Rightarrow \quad v = -\cos x.$$

Substituting in

$$\int u\frac{dv}{dx}\,dx = uv - \int v\frac{du}{dx}\,dx$$

gives

$$\int x^2\sin x\,dx = -x^2\cos x - \int -2x\cos x\,dx$$
$$= -x^2\cos x + \int 2x\cos x\,dx. \qquad ①$$

Now the integral of $2x\cos x$ cannot be found without using integration by parts again. It has to be split into the functions $2x$ and $\cos x$, and as $2x$ becomes a simpler function after differentiation, take $u$ to be $2x$.

$$u = 2x \quad \Rightarrow \quad \frac{du}{dx} = 2$$

$$\frac{dv}{dx} = \cos x \quad \Rightarrow \quad v = \sin x.$$

Substituting in

$$\int u\frac{dv}{dx}\,dx = uv - \int v\frac{du}{dx}\,dx$$

gives

$$\begin{aligned}\int 2x\cos x\,dx &= 2x\sin x - \int 2\sin x\,dx \\ &= 2x\sin x - (-2\cos x) + c \\ &= 2x\sin x + 2\cos x + c.\end{aligned}$$

So in ① $\int x^2\sin x\,dx = -x^2\cos x + 2x\sin x + 2\cos x + c.$

In some cases, the choices of $u$ and $v$ may be less obvious.

**EXAMPLE 3.19**

Find $\int x\ln x\,dx$.

**SOLUTION**

It might seem at first that $u$ should be taken as $x$, because it becomes a simpler function after differentiation.

$$u = x \quad \Rightarrow \quad \frac{du}{dx} = 1$$

$$\frac{dv}{dx} = \ln x.$$

Now you need to integrate $\ln x$ to obtain $v$. Although it is possible to integrate $\ln x$, it has to be done by parts, as you will see in Exercise 3D, question 13. The wrong choice has been made for $u$ and $v$, resulting in a more complicated integral.

So instead, let $u = \ln x$.

$$u = \ln x \quad \Rightarrow \quad \frac{du}{dx} = \frac{1}{x}$$

$$\frac{dv}{dx} = x \quad \Rightarrow \quad v = \tfrac{1}{2}x^2.$$

Substituting in

$$\int u\frac{dv}{dx}\,dx = uv - \int v\frac{du}{dx}\,dx$$

gives

$$\int x\ln x\,dx = \tfrac{1}{2}x^2\ln x - \int \frac{\tfrac{1}{2}x^2}{x}\,dx$$

$$= \tfrac{1}{2}x^2\ln x - \int \tfrac{1}{2}x\,dx$$

$$= \tfrac{1}{2}x^2\ln x - \tfrac{1}{4}x^2 + c.$$

The technique of integration by parts is usually used when the two functions are of different types: polynomials, trigonometric functions, exponentials, logarithms. There are, however, some exceptions, as in questions 11 and 12 of Exercise 3D.

Integration by parts is a very important technique which is needed in many other branches of mathematics. For example, integrals of the form $\int x\,f(x)\,dx$ are used in *Statistics 3* to find the mean for a given probability density function, and in *Mechanics 3* to find the centre of mass for a given shape. Integrals of the form $\int x^2\,f(x)\,dx$, requiring integration by parts to be used twice, are used in *Statistics 3* in finding variance and in *Mechanics 6* to find moments of inertia.

## EXERCISE 3D

*In questions* 1 *to* 6:

**(i)** write down the function to be taken as $u$ and the function to be taken as $\frac{dv}{dx}$;

**(ii)** use the formula for integration by parts to complete the integration.

**1** $\int xe^x\,dx$

**2** $\int x\cos 3x\,dx$

**3** $\int (2x+1)\cos x\,dx$

**4** $\int xe^{-2x}\,dx$

**5** $\int xe^{-x}\,dx$

**6** $\int x\sin 2x\,dx$

*In questions* 7 *to* 10, *use integration by parts to integrate the functions given.*

**7** $x^3\ln x$

**8** $x^2e^x$

**9** $(2-x)^2\cos x$

**10** $x^2\ln 2x$

**11** Find $\int x\sqrt{1+x}\,dx$

**(i)** by using integration by parts;

**(ii)** by using the substitution $u = 1 + x$.

**12** Find $\int 2x(x-2)^4\,dx$

**(i)** by using integration by parts;

**(ii)** by using the substitution $u = x - 2$.

**13 (i)** By writing $\ln x$ as the product of $\ln x$ and 1, use integration by parts to find $\int \ln x\,dx$.

**(ii)** Use the same method to find $\int \ln 3x\,dx$.

## Definite integration by parts

When you use the method of integration by parts on a definite integral, it is important to remember that the term $uv$ on the right-hand side of the expression has already been integrated and so should be written in square brackets with the limits indicated.

$$\int_a^b u\frac{dv}{dx}\,dx = \left[uv\right]_a^b - \int_a^b v\frac{du}{dx}\,dx.$$

**EXAMPLE 3.20**

Evaluate $\int_0^2 xe^x\,dx$.

**SOLUTION**

Put $u = x \quad \Rightarrow \quad \frac{du}{dx} = 1$

and $\frac{dv}{dx} = e^x \quad \Rightarrow \quad v = e^x.$

Substituting in

$$\int_a^b u\frac{dv}{dx}\,dx = \left[uv\right]_a^b - \int_a^b v\frac{du}{dx}\,dx$$

gives

$$\begin{aligned}\int_0^2 xe^x\,dx &= \left[xe^x\right]_0^2 - \int_0^2 e^x\,dx\\ &= [xe^x]_0^2 - [e^x]_0^2\\ &= (2e^2 - 0) - (e^2 - e^0)\\ &= 2e^2 - e^2 + 1\\ &= e^2 + 1.\end{aligned}$$

**EXAMPLE 3.21**

Find the area in the region between the curve $y = x\cos x$ and the $x$ axis, between $x = 0$ and $x = \frac{\pi}{2}$.

**SOLUTION**

Figure 3.9 shows the region whose area is to be found.

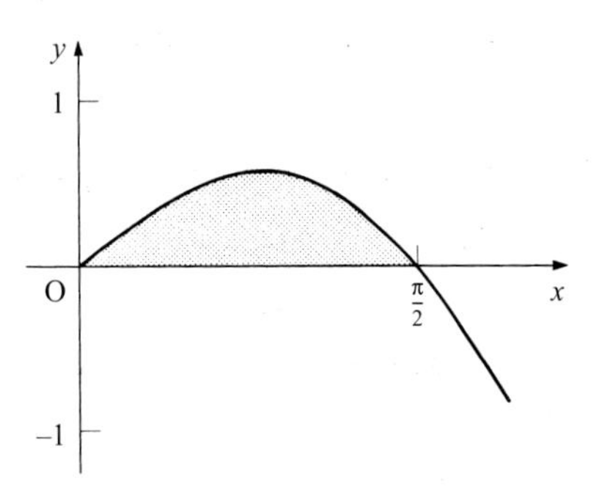


**Figure 3.9**

To find the required area, you need to integrate the function $x\cos x$ between the limits 0 and $\frac{\pi}{2}$. You therefore need to work out

$$\int_0^{\frac{\pi}{2}} x\cos x \, dx.$$

Put $\quad u = x \quad \Rightarrow \quad \frac{du}{dx} = 1$

and $\quad \frac{dv}{dx} = \cos x \quad \Rightarrow \quad v = \sin x$

$$\int_a^b u\frac{dv}{dx}\,dx = \Big[uv\Big]_a^b - \int_a^b v\frac{du}{dx}\,dx.$$

$$\begin{aligned}
\int_0^{\frac{\pi}{2}} x\cos x\,dx &= \Big[x\sin x\Big]_0^{\frac{\pi}{2}} - \int_0^{\frac{\pi}{2}} \sin x\,dx \\
&= \Big[x\sin x\Big]_0^{\frac{\pi}{2}} - \Big[-\cos x\Big]_0^{\frac{\pi}{2}} \\
&= \Big[x\sin x + \cos x\Big]_0^{\frac{\pi}{2}} \\
&= \left(\tfrac{\pi}{2} + 0\right) - (0 + 1) \\
&= \tfrac{\pi}{2} - 1.
\end{aligned}$$

So the required area is $\frac{\pi}{2} - 1$ square units.

## EXERCISE 3E

*Evaluate the definite integrals in questions* 1 *to* 6.

**1** $\int_0^1 xe^{3x}\,dx$

**2** $\int_0^{\pi} (x-1)\cos x\,dx$

**3** $\int_0^2 (x+1)e^x\,dx$

**4** $\int_1^2 \ln 2x\,dx$

**5** $\int_0^{\frac{\pi}{2}} x^2 \sin 2x\,dx$

**6** $\int_1^4 x^2 \ln x\,dx$

**7** **(i)** Find the co-ordinates of the points where the graph of $y = (2-x)e^{-x}$ cuts the $x$ and $y$ axes.

**(ii)** Hence sketch the graph of $y = (2-x)e^{-x}$.

**(iii)** Use integration by parts to find the area of the region between the $x$ axis, the $y$ axis and the graph $y = (2-x)e^{-x}$.

**8** **(i)** Sketch the graph of $y = x\sin x$ from $x = 0$ to $x = \pi$ and shade the region between the curve and the $x$ axis.

**(ii)** Find the area of this region using integration by parts.

**9** Find the area of the region between the $x$ axis, the line $x = 5$ and the graph $y = \ln x$.

**10** Find the area of the region between the $x$ axis and the graph $y = x^2\cos x$ from $x = -\frac{\pi}{2}$ to $x = \frac{\pi}{2}$.

**11** Find the area of the region between the negative $x$ axis and the graph $y = x\sqrt{x+1}$

**(i)** using integration by parts;

**(ii)** using the substitution $u = x + 1$.

**12** The sketch shows the curve with equation $y = x^2\ln 2x$.

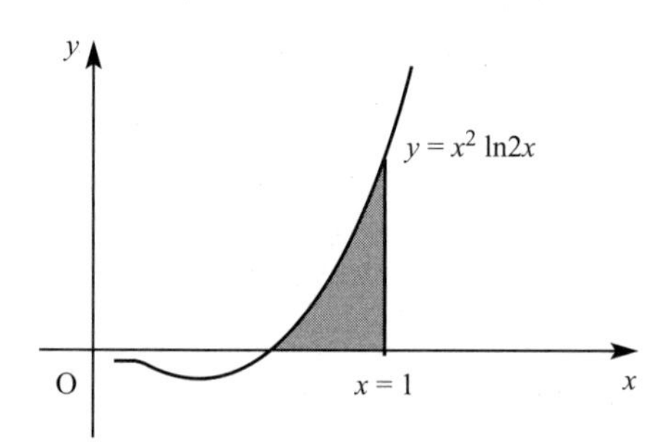


Find the $x$ co-ordinate of the point where the curve cuts the $x$ axis. Hence calculate the area of the shaded region using the method of integration by parts applied to the product of $\ln 2x$ and $x^2$. Give your answer correct to 3 decimal places.

[MEI]

**13 (i)** Use integration by parts to show that

$$\int_1^k t\ln t \, dt = \tfrac{1}{2}k^2 \ln k - \tfrac{1}{4}k^2 + \tfrac{1}{4}$$

where $k$ is positive.

**(ii)** Expand $(1-2x)^{-\frac{1}{2}}$ in ascending powers of $x$, up to and including the term in $x^3$, giving your answer in simplified form. State the range of values of $x$ for which the expansion is valid.

**(iii)** Hence show that, provided $x$ is small, $(1-2x)^{-\frac{1}{2}}\ln(1+x)$ is approximately equal to $t\ln t$, where $t = 1 + x$.

Hence find an approximate value for

$$\int_0^{0.1} \frac{\ln(1+x)}{\sqrt{1-2x}} \, dx.$$

[MEI]

**14 (i)** Find $\int x\cos kx \, dx$, where $k$ is a non-zero constant.

**(ii)** Show that

$$\cos(A-B) - \cos(A+B) = 2\sin A \sin B.$$

Hence express $2\sin 5x \sin 3x$ as the difference of two cosines.

**(iii)** Use the results in parts (i) and (ii) to show that

$$\int_0^{\frac{\pi}{4}} x\sin 5x \sin 3x \, dx = \frac{\pi - 2}{16}.$$

[MEI]

**15 (i)** Find $\int \theta\cos 2\theta \, d\theta$.

**(ii)** Find the expansion of $(1+2x)^{-3}$ up to and including the term in $x^3$, giving the coefficients in their simplest form. State the range of values of $x$ for which the expansion is valid.

**(iii)** Show that, provided $\theta$ is small, $\cos 2\theta \approx 1 - 2\theta^2$.

Hence find $a$ and $b$ such that $(\cos 2\theta)^{-3}$ is approximately $a + b\theta^2$, provided $\theta$ is small.

Use this approximation to find an estimate for

$$\int_0^{0.1} \frac{\theta}{(\cos 2\theta)^3} \, d\theta.$$

[MEI]

**16** Show that $\int_0^1 x^2 e^x \, dx = e - 2$.

Show that the use of the trapezium rule with five strips (six ordinates) gives an estimate that is about 3.8% too high. Explain why approximate evaluation of this integral using the trapezium rule will always result in an overestimate, however many strips are used.

[MEI]

**17** If $I_n = \int_0^1 t^n e^{-t} \, dt$, where $n$ is an integer, show that $I_0 = 1 - e^{-1}$.

By integrating by parts, show that $I_n = nI_{n-1} - e^{-1}$ for $n \geqslant 1$. Hence evaluate $I_3$, leaving your answer in terms of $e^{-1}$.

[MEI]

# The use of partial fractions in integration

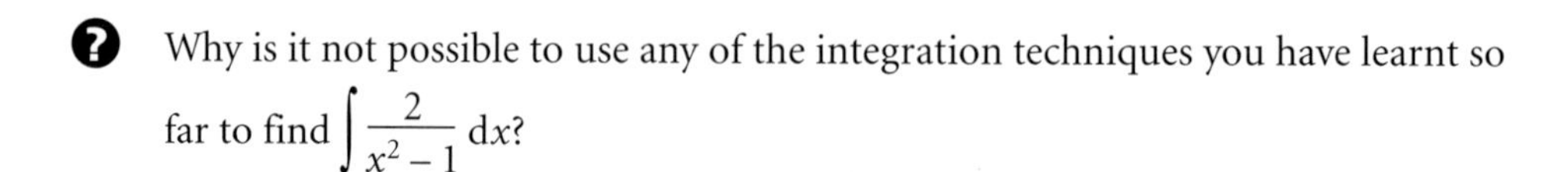

Why is it not possible to use any of the integration techniques you have learnt so far to find $\int \frac{2}{x^2 - 1}\,dx$?

## Partial fractions

Since $x^2 - 1$ can be factorised to give $(x + 1)(x - 1)$, we can put the function to be integrated into partial fractions:

$$\frac{2}{x^2 - 1} = \frac{A}{x - 1} + \frac{B}{x + 1}$$

$$2 \equiv A(x + 1) + B(x - 1).$$

This is true for all values of $x$. It is an identity and to emphasise this point we use the identity symbol $\equiv$

Let $x = 1$ $\qquad 2 = 2A \qquad \Rightarrow \qquad A = 1.$

Let $x = -1$ $\qquad 2 = -2B \qquad \Rightarrow \qquad B = -1.$

Substituting these values for $A$ and $B$ gives

$$\frac{2}{x^2 - 1} = \frac{1}{x - 1} - \frac{1}{x + 1}.$$

The integral then becomes

$$\int \frac{2}{x^2 - 1}\,dx = \int \frac{1}{x - 1}\,dx - \int \frac{1}{x + 1}\,dx.$$

Now the two integrals on the right can be recognised as logarithms. So

$$\int \frac{2}{x^2 - 1}\,dx = \ln|x - 1| - \ln|x + 1| + c$$

$$= \ln\left|\frac{x - 1}{x + 1}\right| + c.$$

Above you worked with the simplest type of partial fraction, in which there are two different linear factors in the denominator. This type will always result in two functions both of which can be integrated to give logarithmic functions. You will now look at the other types of partial fraction.

## A REPEATED FACTOR IN THE DENOMINATOR

**EXAMPLE 3.22**

Find $\int \frac{x+4}{(2x-1)(x+1)^2}\,dx$.

**SOLUTION**

First put the expression into partial fractions:

$$\frac{x+4}{(2x-1)(x+1)^2} = \frac{A}{(2x-1)} + \frac{B}{(x+1)} + \frac{C}{(x+1)^2}$$

where $x+4 \equiv A(x+1)^2 + B(2x-1)(x+1) + C(2x-1)$.

Let $x=-1$ $\quad 3=-3C \Rightarrow C=-1.$

Let $x=\frac{1}{2}$ $\quad \frac{9}{2}=A(\frac{3}{2})^2 \Rightarrow \frac{9}{2}=\frac{9}{4}A \Rightarrow A=2.$

Let $x=0$ $\quad 4=A-B-C \Rightarrow B=A-C-4=2+1-4=-1.$

Substituting these values for $A$, $B$ and $C$ gives

$$\frac{x+4}{(2x-1)(x+1)^2} = \frac{2}{(2x-1)} - \frac{1}{(x+1)} - \frac{1}{(x+1)^2}.$$

Now that the function is in partial fractions, each part can be integrated separately.

$$\int \frac{x+4}{(2x-1)(x+1)^2}\,dx = \int \frac{2}{(2x-1)}\,dx - \int \frac{1}{(x+1)}\,dx - \int \frac{1}{(x+1)^2}\,dx.$$

The first two integrals give logarithmic functions as we saw above. The third, however, is of the form $u^{-2}$ and therefore can be integrated by using the substitution $u=x+1$, or by inspection (i.e. in your head). So

$$\int \frac{x+4}{(2x-1)(x+1)^2}\,dx = \ln|2x-1| - \ln|x+1| + \frac{1}{x+1} + c$$

$$= \ln\left|\frac{2x-1}{x+1}\right| + \frac{1}{x+1} + c.$$

## A QUADRATIC FACTOR IN THE DENOMINATOR

**EXAMPLE 3.23**

Find $\int \frac{x-2}{(x^2+2)(x+1)}\,dx$.

**SOLUTION**

First put the expression into partial fractions:

$$\frac{x-2}{(x^2+2)(x+1)} = \frac{Ax+B}{(x^2+2)} + \frac{C}{(x+1)}$$

where $x-2 \equiv (Ax+B)(x+1) + C(x^2+2)$.

Rearranging gives

$$x - 2 \equiv (A + C)x^2 + (A + B)x + (B + 2C).$$

Equating coefficients:

$$x^2 \quad \Rightarrow \quad A + C = 0$$

$$x \quad \Rightarrow \quad A + B = 1$$

$$\text{constant terms} \quad \Rightarrow \quad B + 2C = -2.$$

Solving these gives $A = 1$, $B = 0$, $C = -1$. Hence

$$\frac{x-2}{(x^2+2)(x+1)} = \frac{x}{(x^2+2)} - \frac{1}{(x+1)}$$

$$\int \frac{x-2}{(x^2+2)(x+1)}\,dx = \int \frac{x}{(x^2+2)}\,dx - \int \frac{1}{(x+1)}\,dx$$

$$= \frac{1}{2}\int \frac{2x}{x^2+2}\,dx - \int \frac{1}{x+1}\,dx$$

$\frac{1}{2}\ln|x^2+2| = \ln\sqrt{x^2+2}$
Notice that $(x^2 + 2)$ is positive for all values of $x$

$$= \tfrac{1}{2}\ln|x^2+2| - \ln|x+1| + c$$

$$= \ln\left|\frac{\sqrt{x^2+2}}{x+1}\right| + c.$$

### *Note*

If $B$ had not been 0, you would have had an expression of the form $\dfrac{Ax+B}{x^2+2}$ to integrate. This can be split into

$$\frac{Ax}{x^2+2} + \frac{B}{x^2+2}$$

The first part of this can be integrated as in Example 3.23, but the second part cannot be integrated by any method you have met so far. If you go on to study Pure Mathematics 5, you will meet integrals of this form then. If in the meantime you come across a case (e.g. in modelling a situation) where you need to find such an integral, you may choose to use the standard result that

$$\int \frac{1}{(x^2+a^2)}\,dx = \frac{1}{a}\arctan\left(\frac{x}{a}\right) + c.$$

## EXERCISE 3F



**1** Express the functions in each of the following integrals in partial fractions, and hence perform the integration.

**(i)** $\displaystyle\int \frac{1}{(1-x)(3x-2)}\,dx$ **(ii)** $\displaystyle\int \frac{7x-2}{(x-1)^2(2x+3)}\,dx$

**(iii)** $\displaystyle\int \frac{x+1}{(x^2+1)(x-1)}\,dx$ **(iv)** $\displaystyle\int \frac{3x+3}{(x-1)(2x+1)}\,dx$

**(v)** $\displaystyle\int \frac{1}{x^2(1-x)}\,dx$ **(vi)** $\displaystyle\int \frac{1}{(x+1)(x+3)}\,dx$

**(vii)** $\displaystyle\int \frac{2x-4}{(x^2+4)(x+2)}\,dx$ **(viii)** $\displaystyle\int \frac{5x+1}{(x+2)(2x+1)^2}\,dx$

**2** Express in partial fractions the function

$$f(x) = \frac{3x+4}{(x^2+4)(x-3)}$$

and hence find $\int_0^2 f(x)\,dx$.

[MEI]

**3** Express $\dfrac{1}{x^2(2x+1)}$ in partial fractions. Hence show that

$$\int_1^2 \frac{dx}{x^2(2x+1)} = \tfrac{1}{2} + 2\ln\tfrac{5}{6}.$$

[MEI]

**4** **(i)** **(a)** Express $\dfrac{3}{(1+x)(1-2x)}$ in partial fractions.

**(b)** Hence find

$$\int_0^{0.1} \frac{3}{(1+x)(1-2x)}\,dx$$

giving your answer to 5 decimal places.

**(ii)** **(a)** Find the first three terms in the binomial expansion of $3(1+x)^{-1}(1-2x)^{-1}$.

**(b)** Use the first three terms of this expansion to find an approximation for

$$\int_0^{0.1} \frac{3}{(1+x)(1-2x)}\,dx$$

**(c)** What is the percentage error in your answer to part (b)?

**5 (i)** Given that

$$\frac{x^2 - x - 24}{(x+2)(x-4)} \equiv A + \frac{B}{(x+2)} + \frac{C}{(x-4)},$$

find the values of the constants $A$, $B$ and $C$.

**(ii)** Find $\int_1^3 \frac{x^2 - x - 24}{(x+2)(x-4)}\,\mathrm{d}x$.

[MEI]

**6 (i)** Find $\int x\mathrm{e}^{2x}\,\mathrm{d}x$.

**(ii)** Find the exact value of $\int_{\frac{1}{4}\pi}^{\pi} \sin^2 3x\,\mathrm{d}x$.

**(iii)** The expression $\frac{x^2}{(x-4)^2(x-2)}$ is to be written in partial fractions of the form

$$\frac{A}{(x-4)^2} + \frac{B}{x-4} + \frac{C}{x-2}.$$

Show that $B = 0$ and find $A$ and $C$.

Hence show that $\int_5^8 \frac{x^2}{(x-4)^2(x-2)}\,\mathrm{d}x = 6 + \ln 2$.

[MEI]

**7 (i)** Express the function

$$\mathrm{f}(x) = \frac{1 - 3x}{(1+2x)(1+x^2)}$$

in the form $\frac{A}{1+2x} + \frac{Bx + C}{1+x^2}$.

**(ii)** Use the binomial series to show that, for suitably small values of $x$,

$$\mathrm{f}(x) \approx 1 - 5x + 9x^2.$$

State the range of values of $x$ for which the binomial series expansion is valid.

**(iii)** By using a small-angle approximation for $\sin\theta$, together with the result in part (ii) above, find an approximation for

$$\int_0^{0.1} \frac{1 - 3\sin\theta}{(1 + 2\sin\theta)(1+\theta^2)}\,\mathrm{d}\theta.$$

[MEI]

**8 (i)** Given that

$$\mathrm{f}(x) = \frac{16 + 2x + 15x^2}{(1+x^2)(2-x)} \equiv \frac{A + Bx}{1+x^2} + \frac{C}{2-x},$$

find the values of $B$ and $C$ and show that $A = 0$.

**(ii)** Find $\int_0^1 \mathrm{f}(x)\,\mathrm{d}x$ in an exact form.

**(iii)** Express $\mathrm{f}(x)$ as a sum of powers of $x$ up to and including the term in $x^4$. Determine the range of values of $x$ for which this expansion of $\mathrm{f}(x)$ is valid.

[MEI]

# General integration



You now know several techniques for integration which can be used to integrate a wide variety of functions. One of the difficulties which you may now experience when faced with an integration is deciding which technique is appropriate! This section gives you some guidelines on this, as well as revising all the work on integration that you have done so far.

---

**?** Look at the integrals below and try to decide which technique you would use and, in the case of a substitution, what function you would write as $u$. Do not attempt actually to carry out the integrations. Make a note of your decisions – you will return to these integrals later.

**(i)** $\displaystyle\int \frac{x-5}{x^2+2x-3}\,dx$ **(ii)** $\displaystyle\int \frac{x+1}{x^2+2x-3}\,dx$

**(iii)** $\int xe^x\,dx$ **(iv)** $\int xe^{x^2}\,dx$

**(v)** $\int \sin^2 x\,dx$ **(vi)** $\int \cos x \sin^2 x\,dx$

---

## Choosing an appropriate method of integration

You have now met the following standard integrals.

| f($x$) | $\int$f($x$) d$x$ |
|---|---|
| $x^n \ (n \neq -1)$ | $\dfrac{x^{n+1}}{n+1}$ |
| $\dfrac{1}{x}$ | $\ln \lvert x \rvert$ |
| $e^x$ | $e^x$ |
| $\sin x$ | $-\cos x$ |
| $\cos x$ | $\sin x$ |

If you are asked to integrate any of these standard functions, you may simply write down the answer.

For other integrations, the following table may help.

| Type of function to be integrated | Examples | Method of integration |
|---|---|---|
| Simple variations of any of the standard functions | $\cos(2x+1)$<br>$e^{3x}$ | Substitution may be used, but it should be possible to do these by inspection |
| Product of two functions of the form $f'(x)g[f(x)]$<br>Note that $f'(x)$ means $\frac{d}{dx}[f(x)]$ | $2xe^{x^2}$<br>$x^2(x^3+1)^6$ | Substitution $u = f(x)$ |
| Other products, particularly when one function is a small positive integral power of $x$ or a polynomial in $x$ | $xe^x$<br>$x^2\sin x$ | Integration by parts |
| Quotients of the form $\frac{f'(x)}{f(x)}$ or functions which can easily be converted to this form | $\frac{x}{x^2+1}$<br>$\frac{\sin x}{\cos x}$ | Substitution $u = f(x)$, or better by inspection: $k\ln\lvert f(x)\rvert + c$, where $k$ is known |
| Polynomial quotients which may be split into partial fractions | $\frac{x+1}{x(x-1)}$<br>$\frac{x-4}{x^2-x-2}$ | Split into partial fractions and integrate term by term |
| Even powers of $\sin x$ or $\cos x$ | $\sin^2 x$<br>$\cos^4 x$ | Use the double-angle formulae to transform the function before integrating |
| Odd powers of $\sin x$ or $\cos x$ | $\cos^3 x$ | Use $\cos^2 x + \sin^2 x = 1$ and write in form $f'(x)g[f(x)]$ |

It is impossible to give an exhaustive list of possible types of integration, but the above table and that on the previous page cover the most common situations that you will meet.

**ACTIVITY**

Now look back at the integrals in the discussion point at the beginning of this section and the decisions you made about which method of integration should be used for each one.

**(i)** $\int \frac{x-5}{x^2+2x-3}\,dx$ **(ii)** $\int \frac{x+1}{x^2+2x-3}\,dx$

**(iii)** $\int xe^x\,dx$ **(iv)** $\int xe^{x^2}\,dx$

**(v)** $\int \sin^2 x\,dx$ **(vi)** $\int \cos x\sin^2 x\,dx$

**(i)** This is a quotient. The derivative of the function on the bottom is not related to the function on the top, so you cannot use substitution. However, as the function on the bottom can be factorised, you can put it into partial fractions.

**(ii)** The derivative of the function on the bottom line is $2x + 2$, which is twice the function on the top line. So the integral is of the form

$$k\int \frac{\mathrm{f}'(x)}{\mathrm{f}(x)}\,\mathrm{d}x = k\ln|\mathrm{f}(x)| + c.$$

This integral can also be found using partial fractions, but using logarithms is quicker.

**(iii)** This is a product of $x$ and $\mathrm{e}^x$. There is no relationship between one function and the derivative of the other, so you cannot use substitution. As one of the functions is $x$, you can use integration by parts.

**(iv)** This is also a product, this time of $x$ and $\mathrm{e}^{x^2}$. $\mathrm{e}^{x^2}$ is a function of $x^2$, and $2x$ is the derivative of $x^2$, so you can use the substitution $u = x^2$.

**(v)** This is one of the special cases where you need to use a trigonometric identity. Use the identity $\sin^2 x = \frac{1}{2}(1 - \cos 2x)$.

**(vi)** This is a product: $\sin^2 x$ is a function of $\sin x$, and $\cos x$ is the derivative of $\sin x$, so you can use the substitution $u = \sin x$.

Check if your original ideas were correct. Then integrate the six functions you have just been looking at, using the suggestions given above.

## EXERCISE 3G

**1** Choose an appropriate method and integrate the following functions. You may find it helpful first to discuss in class which method to use.

**(i)** $\int \cos(3x - 1)\,\mathrm{d}x$

**(ii)** $\int \frac{2x + 1}{(x^2 + x - 1)^2}\,\mathrm{d}x$

**(iii)** $\int \sec^2 x \tan^2 x\,\mathrm{d}x$

**(iv)** $\int \mathrm{e}^{1-x}\,\mathrm{d}x$

**(v)** $\int x^2 \sin 2x\,\mathrm{d}x$

**(vi)** $\int \cos^2 x\,\mathrm{d}x$

**(vii)** $\int \ln 2x\,\mathrm{d}x$

**(viii)** $\int \frac{x}{(x^2 - 1)^3}\,\mathrm{d}x$

**(ix)** $\int \sqrt{2x - 3}\,\mathrm{d}x$

**(x)** $\int \frac{4x - 1}{(x - 1)^2(x + 2)}\,\mathrm{d}x$

**(xi)** $\int \sin^3 2x\,\mathrm{d}x$

**(xii)** $\int x^3 \ln x\,\mathrm{d}x$

**(xiii)** $\int \frac{5}{2x^2 - 7x + 3}\,\mathrm{d}x$

**(xiv)** $\int (x + 1)\mathrm{e}^{x^2 + 2x}\,\mathrm{d}x$

**2** Evaluate the following definite integrals.

**(i)** $\int_8^{24} \frac{dx}{\sqrt{3x-8}}$ **(ii)** $\int_8^{24} \frac{dx}{3x-8}$ **(iii)** $\int_8^{24} \frac{9x}{3x-8}\,dx$

**(iv)** $\int_0^{\frac{\pi}{3}} \sin^3 x\,dx$ **(v)** $\int_1^2 x^2 \ln x\,dx$

**3** Evaluate $\int_8^2 \frac{x^2}{\sqrt{1+x^3}}\,dx$, using the substitution $u = 1 + x^3$, or otherwise.

[MEI]

**4** Find $\int_0^{\frac{\pi}{4}} \frac{\sin\theta}{\cos^4\theta}\,d\theta$ in terms of $\sqrt{2}$.

[MEI]

**5 (i)** Find $\int_0^{\frac{\pi}{2}} \sin^2\theta\,d\theta$, leaving your answer in terms of $\pi$.

**(ii)** Using the substitution $u = \ln x$, or otherwise, find $\int_1^2 \frac{\ln x}{x}\,dx$, giving your answer to 2 decimal places.

[MEI]

**6** Find $\int_0^{\frac{\pi}{4}} x\cos 2x\,dx$, expressing your answer in terms of $\pi$.

[MEI]

**7 (i)** Find $\int xe^{-2x}\,dx$.

**(ii)** Evaluate $\int_0^1 \frac{x}{(4+x^2)}\,dx$, giving your answer correct to 3 significant figures.

[MEI]

**8 (i)** Find $\int \sin(2x-3)\,dx$.

**(ii)** Use the method of integration by parts to evaluate $\int_0^2 xe^{2x}\,dx$.

**(iii)** Using the substitution $t = x^2 - 9$, or otherwise, find $\int \frac{x}{x^2-9}\,dx$.

[MEI]

**9** Evaluate

**(i)** $\int_0^1 (2x^2+1)(2x^3+3x+4)^{\frac{1}{2}}\,dx$

**(ii)** $\int_0^e \frac{\ln x}{x^3}\,dx$

[MEI]

**10** Find $\int_0^{\frac{\pi}{2}} \sin x\cos^3 x\,dx$ and $\int_0^1 t\,e^{-2t}\,dt$.

[MEI]

## KEY POINTS

**1** $\dfrac{d}{dx}(\sin kx) = k\cos kx$

$\dfrac{d}{dx}(\cos kx) = -k\sin kx$

$\dfrac{d}{dx}(\tan kx) = k\sec^2 kx$

**2** $\displaystyle\int \sin kx \, dx = -\frac{1}{k}\cos kx + c$

$\displaystyle\int \cos kx \, dx = -\frac{1}{k}\sin kx + c$

**3** Even powers of $\sin x$ and $\cos x$ can be integrated by using the identities

$\sin^2 x = \frac{1}{2}(1 - \cos 2x)$ and $\cos^2 x = \frac{1}{2}(1 + \cos 2x)$.

**4** Odd powers of $\sin x$ and $\cos x$ can be integrated after using the identity

$\sin^2 x + \cos^2 x = 1$.

**5** Some products may be integrated by parts using the formula

$$\int u\frac{dv}{dx}\,dx = uv - \int v\frac{du}{dx}\,dx.$$

**6** Some functions may be integrated by first splitting them into partial fractions.

**7** $\displaystyle\int \frac{f'(x)}{f(x)}\,dx = \ln|f(x)| + c$

# 4 Parametric co-ordinates

**A mathematician, like a painter or poet, is a maker of patterns. If his patterns are more permanent than theirs it is because they are made with ideas.**

*G.H. Hardy*

When you go on a ride like the one in the picture, your body follows a very unnatural path and this gives rise to sensations which you may find exhilarating or frightening.

You are accustomed to expressing curves as mathematical equations. How would you do so in a case like this?

Figure 4.1 shows a simplified version of such a ride.

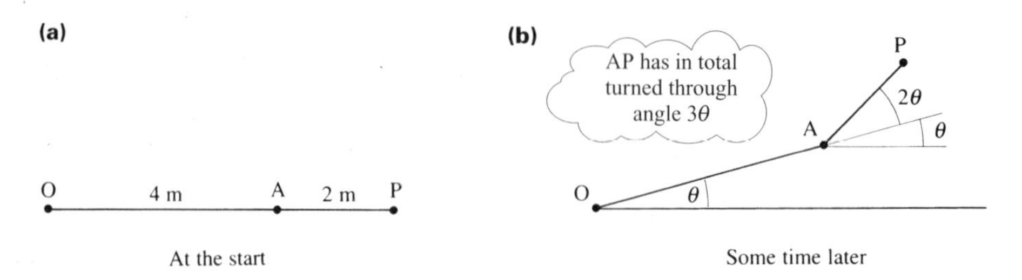


**Figure 4.1**

The passenger's chair is on the end of a rod AP of length 2 m which is rotating about A. The rod OA is 4 m long and is itself rotating about O. The gearing of the mechanism ensures that the rod AP rotates twice as fast relative to OA as the rod OA does. This is illustrated by the angles marked on figure 4.1(b), at a time when OA has rotated through an angle $\theta$.

At this time, the co-ordinates of the point P, taking O as the origin, are given by

$$x = 4\cos\theta + 2\cos3\theta$$

$$y = 4\sin\theta + 2\sin3\theta$$

(see figure 4.2).

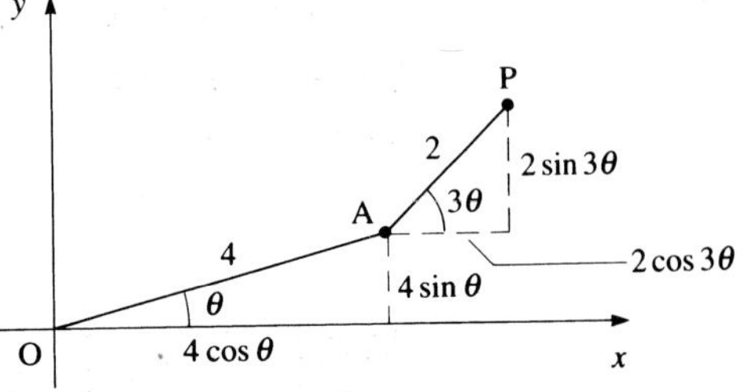


**Figure 4.2**

These two equations are called *parametric equations* of the curve. They do not give the relationship between $x$ and $y$ directly in the form $y = f(x)$ but use a third variable, $\theta$, to do so. This third variable is called the *parameter*.

To plot the curve, you need to substitute values of $\theta$ and find the corresponding values of $x$ and $y$.

Thus $\theta = 0° \Rightarrow x = 4 + 2 = 6$

$y = 0 + 0 = 0$ Point (6, 0)

$\theta = 30° \Rightarrow x = 4 \times 0.866 + 0 = 3.464$

$y = 4 \times 0.5 + 2 \times 1 = 4$ Point (3.46, 4)

and so on.

Joining points found in this way reveals the curve to have the shape shown in figure 4.3.

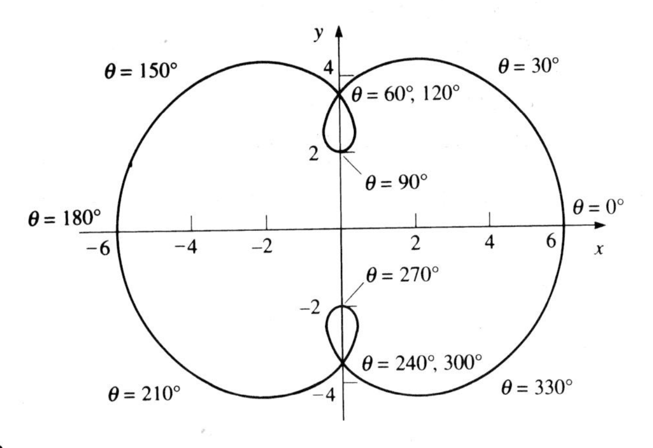


**Figure 4.3**

At what points of the curve would you feel the greatest sensations?

## Graphs from parametric equations

Parametric equations are very useful in situations such as this, where an otherwise complicated equation may be expressed reasonably simply in terms of a parameter. Indeed, there are some curves which can be given by parametric equations but cannot be written as cartesian equations (in terms of $x$ and $y$ only).

The next example is based on a simpler curve. Make sure that you can follow the solution completely before going on to the rest of the chapter.

**EXAMPLE 4.1**

A curve has the parametric equations $x = 2t$, $y = \frac{36}{t^2}$.

**(i)** Find the co-ordinates of the points corresponding to $t = 1, 2, 3, -1, -2$ and $-3$.

**(ii)** Plot the points you have found and join them to give the curve.

**(iii)** Explain what happens as $t \to 0$.

**SOLUTION**

**(i)**

| $t$ | $-3$ | $-2$ | $-1$ | $1$ | $2$ | $3$ |
|---|---|---|---|---|---|---|
| $x$ | $-6$ | $-4$ | $-2$ | $2$ | $4$ | $6$ |
| $y$ | $4$ | $9$ | $36$ | $36$ | $9$ | $4$ |

The points required are $(-6, 4)$, $(-4, 9)$, $(-2, 36)$, $(2, 36)$, $(4, 9)$ and $(6, 4)$.

**(ii)** The curve is shown in figure 4.4.

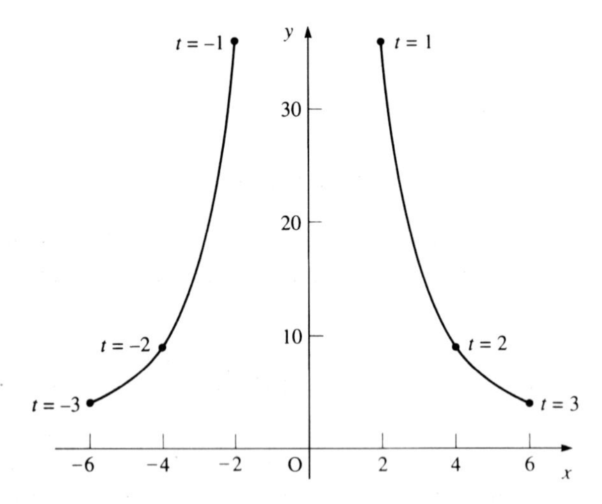


**Figure 4.4**

**(iii)** As $t \to 0$, $x \to 0$ and $y \to \infty$. The $y$ axis is an asymptote for the curve.

**EXAMPLE 4.2**

A curve has the parametric equations $x = t^2$, $y = t^3 - t$.

**(i)** Find the co-ordinates of the points corresponding to values of $t$ from $-2$ to $+2$ at half-unit intervals.

**(ii)** Sketch the curve for $-2 \leqslant t \leqslant 2$.

**(iii)** Are there any values of $x$ for which the curve is undefined?

**SOLUTION**

**(i)**

| $t$ | $-2$ | $-1.5$ | $-1$ | $-0.5$ | $0$ | $0.5$ | $1$ | $1.5$ | $2$ |
|---|---|---|---|---|---|---|---|---|---|
| $x$ | $4$ | $2.25$ | $1$ | $0.25$ | $0$ | $0.25$ | $1$ | $2.25$ | $4$ |
| $y$ | $-6$ | $-1.875$ | $0$ | $0.375$ | $0$ | $-0.375$ | $0$ | $1.875$ | $6$ |

**(ii)**

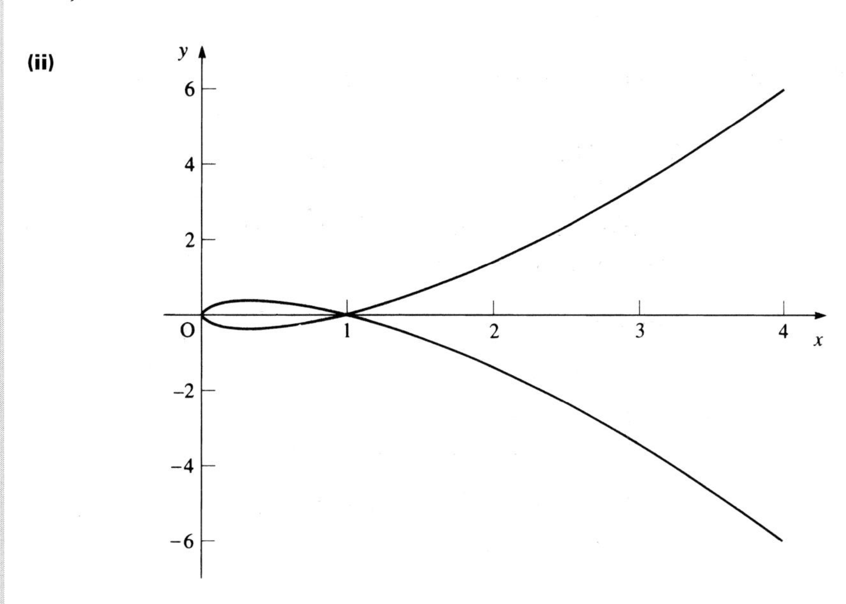


**Figure 4.5**

**(iii)** The curve in figure 4.5 is undefined for $x < 0$.

### Graphics calculators

Graphics calculators can be used to sketch parametric curves, but as with cartesian curves, some insight is needed when choosing the range.

## Finding the equation by eliminating the parameter

For some pairs of parametric equations, it is possible to eliminate the parameter and obtain the cartesian equation for the curve. This is usually done by making the parameter the subject of one of the equations, and substituting this expression into the other.

**EXAMPLE 4.3**

Eliminate $t$ from the equations $x = t^3 - 2t^2$, $y = \frac{t}{2}$.

**SOLUTION**

$$y = \frac{t}{2} \quad \Rightarrow \quad t = 2y.$$

Substituting this in the equation $x = t^3 - 2t^2$ gives

$$x = (2y)^3 - 2(2y)^2 \quad \text{or} \quad x = 8y^3 - 8y^2.$$

Sometimes you need to consider the parametric equations simultaneously. There is often more than one way in which you can do this, and the next example gives two different options.

**EXAMPLE 4.4**

The parametric equations of a curve are

$$x = t + \frac{1}{t} \qquad y = t - \frac{1}{t}.$$

**(i)** Find the co-ordinates of the points corresponding to $t = -2, -1, -0.5, 0, 0.5, 1, 2$.

**(ii)** Sketch the curve for $-2 \leqslant t \leqslant 2$.

**(iii)** For what values of $x$ is the curve undefined?

**(iv)** Eliminate the parameter by

  **(a)** first finding $x + y$

  **(b)** first squaring $x$ and $y$.

**SOLUTION**

**(i)**

| $t$ | $-2$ | $-1$ | $-0.5$ | $0$ | $0.5$ | $1$ | $2$ |
|---|---|---|---|---|---|---|---|
| $x$ | $-2.5$ | $-2$ | $-2.5$ | undefined | $2.5$ | $2$ | $2.5$ |
| $y$ | $-1.5$ | $0$ | $1.5$ | undefined | $-1.5$ | $0$ | $1.5$ |

**(ii)**

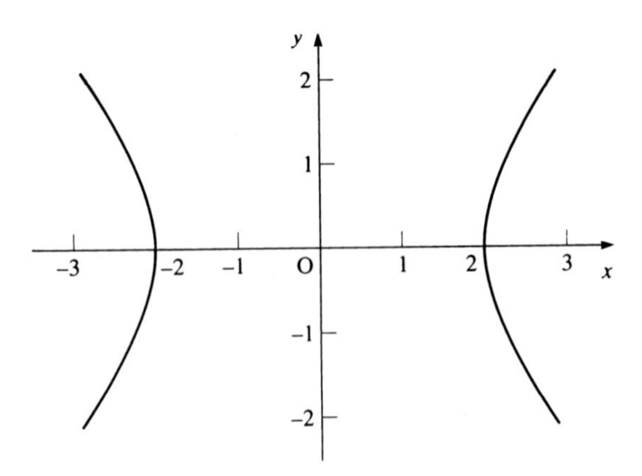


**Figure 4.6**

**(iii)** The curve is undefined for $-2 < x < 2$.

**(iv) (a)** Adding the two equations gives

$$x + y = 2t \qquad \text{or} \qquad t = \frac{x + y}{2}.$$

Substituting for $t$ in the first equation (it could be either one) gives

$$x = \frac{x + y}{2} + \frac{2}{x + y}.$$

At this point the parameter $t$ has been eliminated, but the equation is not in its neatest form. Multiplying by $2(x + y)$ to eliminate the fractions:

$$2x(x + y) = (x + y)^2 + 4$$

$$\Rightarrow \qquad 2x^2 + 2xy = x^2 + 2xy + y^2 + 4$$

$$\Rightarrow \qquad x^2 - y^2 = 4.$$

**(b)** Squaring gives

$$x^2 = t^2 + 2 + \frac{1}{t^2}$$

$$y^2 = t^2 - 2 + \frac{1}{t^2}.$$

Subtracting gives

$$x^2 - y^2 = 4.$$

*Note*

Figure 4.7 shows that the curve is the rectangular hyperbola $xy = 2$ rotated clockwise through 45°.

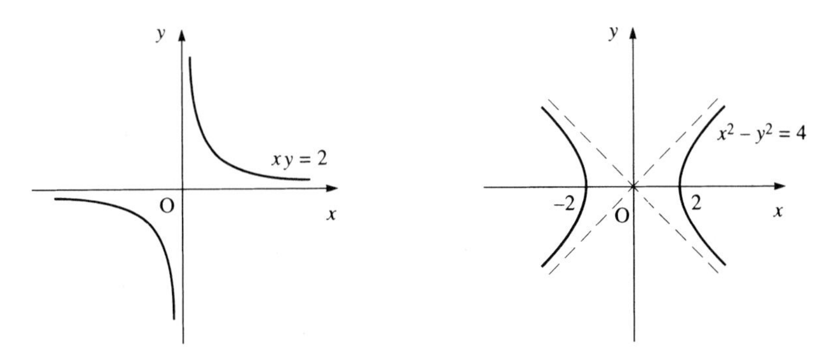


**Figure 4.7**

## Trigonometric parametric equations

When trigonometric functions are used in parametric equations, a particular trigonometric identity may help you to eliminate the parameter. The next example illustrates this.

**EXAMPLE 4.5**

Eliminate $\theta$ from $x = 4\cos\theta$, $y = 3\sin\theta$.

**SOLUTION**

The identity which connects $\cos\theta$ and $\sin\theta$ is

$$\cos^2\theta + \sin^2\theta = 1 \qquad ①$$

$$x = 4\cos\theta \quad \Rightarrow \quad \cos\theta = \frac{x}{4}$$

$$y = 3\sin\theta \quad \Rightarrow \quad \sin 0 = \frac{y}{3}.$$

Substituting these in ① gives

$$\left(\frac{x}{4}\right)^2 + \left(\frac{y}{3}\right)^2 = 1.$$

This is usually written as

$$\frac{x^2}{16} + \frac{y^2}{9} = 1$$

and is the equation of the ellipse shown in figure 4.8.

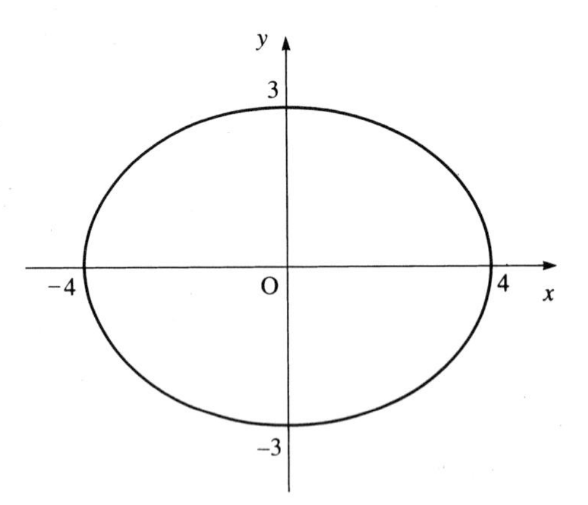


**Figure 4.8**

*Note*

The standard equation of the ellipse is $\frac{x^2}{a^2} + \frac{y^2}{b^2} = 1$ and this crosses the $x$ axis at $(-a, 0)$ and $(a, 0)$ and the $y$ axis at $(0, b)$ and $(0, -b)$.

The expansions of $\cos 2\theta$ in terms of either $\sin\theta$ or $\cos\theta$ are also useful in this context.





**EXAMPLE 4.6**

Eliminate $\theta$ from $x = \cos 2\theta$, $y = \sin\theta + 2$.

**SOLUTION**

The relationship between $\cos 2\theta$ and $\sin\theta$ is

$$\cos 2\theta = 1 - 2\sin^2\theta.$$

Now $\quad y - 2 = \sin\theta$

so $\quad x = 1 - 2(y - 2)^2.$

## Parametric equations of some standard curves

### Circle

The circle with centre (0, 0) and radius 4 units has the equation $x^2 + y^2 = 16$. Alternatively, using the triangle OAB and the angle $\theta$ in figure 4.9, we can write the equations

$$x = 4\cos\theta$$

$$y = 4\sin\theta.$$

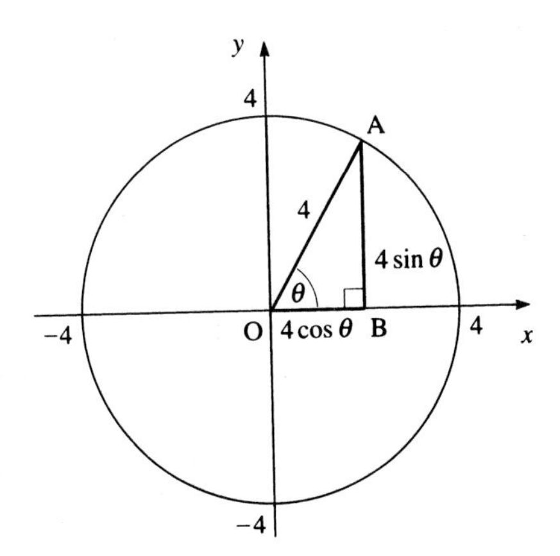


**Figure 4.9**

Generalising, a circle with centre (0, 0) and radius $r$ has the parametric equations

$$x = r\cos\theta$$

$$y = r\sin\theta.$$

Translating the centre of the circle to the point $(a, b)$ gives the circle in figure 4.10 whose parametric equations are

$x = a + r\cos\theta$
$y = b + r\sin\theta$

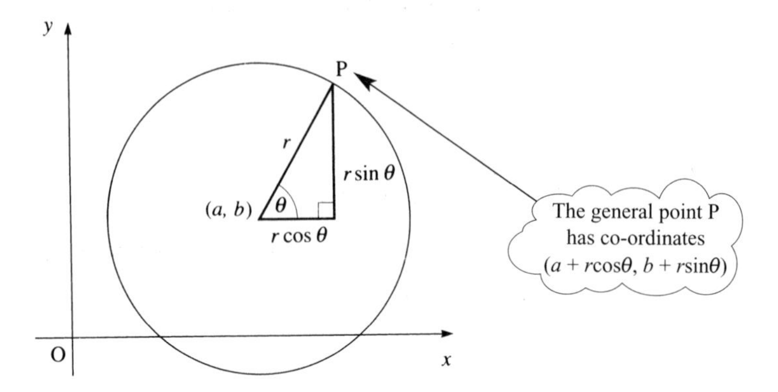


**Figure 4.10**

## Ellipse

In Example 4.5, we saw that the parametric equations

$$x = 4\cos\theta$$

$$y = 3\sin\theta$$

were equivalent to the cartesian equation

$$\frac{x^2}{16} + \frac{y^2}{9} = 1.$$

In general the equations

$$x = a\cos\theta \qquad y = b\sin\theta$$

correspond to the ellipse

$$\frac{x^2}{a^2} + \frac{y^2}{b^2} = 1$$

(see figure 4.11).

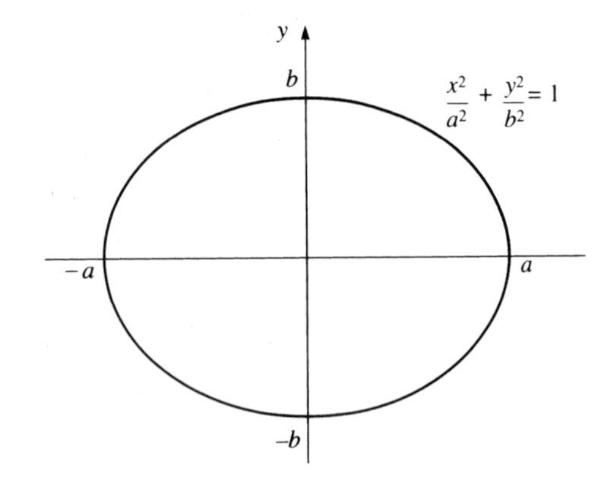


**Figure 4.11**

Here the parameter $\theta$ is not an angle in the ellipse, as it was in the circle. It does, however, have a physical interpretation as an angle in the circumscribing circle.

## Parabola

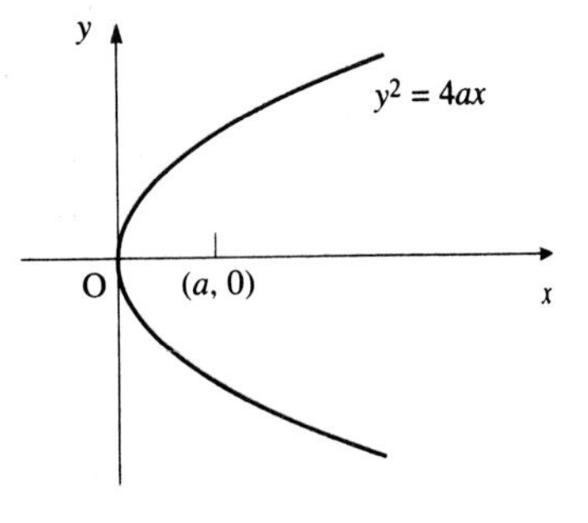


Figure 4.12

The parabola in figure 4.12 whose line of symmetry is the $x$ axis, and whose focus is at the point $(a, 0)$, has the cartesian equation $y^2 = 4ax$. The corresponding parametric equations are

$$x = at^2 \qquad y = 2at.$$

## Rectangular hyperbola

The rectangular hyperbola $xy = c^2$ shown in figure 4.13(a) has the parametric equations

$$x = ct \qquad y = \frac{c}{t}.$$

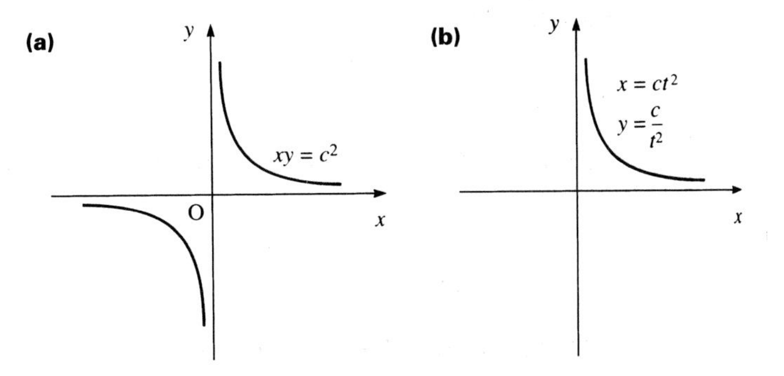


Figure 4.13

---

When you convert the equation of a curve from parametric to cartesian form, you must take care that there are no restrictions on the values of $x$ and $y$. For example, in figure 4.13(b), the curve $x = ct^2$, $y = \frac{c}{t^2}$ $(c > 0)$ is restricted to positive values of $x$ and $y$ (since $t^2 > 0$). However, its cartesian form, $xy = c^2$, would appear to allow negative values of $x$ and $y$.

---

*Note*

When the parametric equations can be recognised as those of a standard curve, the curve can be sketched immediately without the need for further investigation.

## EXERCISE 4A

*In this exercise you should sketch the curves by hand. If you have access to a graphics calculator, you can use it to check your results.*

**1** In each of the following examples.

**(i)** find the co-ordinates of the points corresponding to values of $t$ from $-2$ to $+2$ in half-unit intervals, or values of $\theta$ from $0°$ to $360°$ in $30°$ intervals;

**(ii)** sketch the curve;

**(iii)** find the cartesian equation of the curve.

**(a)** $x = 2t$, $y = t^2$ **(b)** $x = \cos 2\theta$, $y = \sin^2\theta$ **(c)** $x = t^2$, $y = t^3$

**(d)** $x = \sin^2\theta$, $y = 1 + 2\sin\theta$ **(e)** $x = 2\operatorname{cosec}\theta$, $y = 2\cot\theta$ **(f)** $x = 2\sin^2\theta$, $y = 3\cos\theta$

**(g)** $x = \tan\theta$, $y = \tan 2\theta$ **(h)** $x = t^2$, $y = t^2 - t$ **(i)** $x = \dfrac{t}{1+t}$, $y = \dfrac{t}{1-t}$

**2** Sketch the standard curves given by the following equations.

**(i)** $x = 5\cos\theta$, $y = 5\sin\theta$ **(ii)** $x = 3\cos\theta$, $y = 2\sin\theta$

**(iii)** $x = 4 + 3\cos\theta$, $y = 1 + 3\cos\theta$ **(iv)** $x = 2\cos\theta - 1$, $y = 3 + 2\sin\theta$

**3 (i)** Sketch both of these curves on the same axes.

**(a)** $x = t \quad y = \dfrac{1}{t}$ **(b)** $x = 4t \quad y = \dfrac{4}{t}$

**(ii)** Comment on the relationship between them.

**4 (i)** Sketch the curve given by the parametric equations $x = 5\cos\theta$, $y = 3\sin\theta$.

**(ii)** State the parametric equations of the inscribed circle (the largest circle which fits inside the curve) and the circumscribing circle.

**5** A curve has the parametric equations $x = t^2$, $y = t^4$.

**(i)** Find the co-ordinates of the points corresponding to $t = -2$ to $t = 2$ at half-unit intervals.

**(ii)** Sketch the curve for $-2 \leqslant t \leqslant 2$.

**(iii)** Why is it not quite accurate to say this curve has equation $y = x^2$?

**6** When a tennis ball is served in still air, its trajectory (path) may be modelled by the parametric equations $x = 20t$, $y = 10t - 5t^2$, where $t$ is the time in seconds after the service.

**(i)** Find the cartesian equation of its trajectory.

**(ii)** Sketch its trajectory.

**7** A student is investigating the trajectory of a golf ball being hit over level ground. At first she ignores air resistance and this leads her to an initial model given by $x = 40t$, $y = 30t - 5t^2$, where $x$ and $y$ are the horizontal and vertical distances in metres from where the ball is hit, and $t$ is the time in seconds.

**(i)** Plot the trajectory on graph paper for $t = 0, 1, 2, \ldots$, until the ball hits the ground again.

**(ii)** How far does the ball travel horizontally before bouncing, according to this model?

The student then decides to make an allowance for air resistance to the horizontal motion and proposes the model $x = 40t - t^2$, $y = 30t - 5t^2$.

**(iii)** Plot the trajectory according to this model using the same axes as you did in part (i).

**(iv)** By how much does this model reduce the horizontal distance the ball travels before bouncing?

**8** A curve has parametric equations $x = (t + 1)^2$, $y = t - 1$.

**(i)** Find the co-ordinates of the points corresponding to $t = -4$ to $t = 4$ at intervals of one unit.

**(ii)** Sketch the curve for $-4 \leqslant t \leqslant 4$.

**(iii)** State the equation of the line of symmetry of the curve.

**(iv)** By eliminating the parameter, find the cartesian equation of the curve.

**9** A curve has parametric equations $x = e^t$, $y = \sin t$, where $t$ is in radians.

**(i)** Find, to 2 decimal places, the co-ordinates of the points corresponding to values of $t$ from $-2$ to $+2$ at half-unit intervals.

**(ii)** What can you say about the values of $x$ for which the curve is defined?

**(iii)** Sketch the curve for $-2 \leqslant t \leqslant 2$.

**(iv)** Predict how this graph would continue if all values of $t$ were considered (that is, $t < -2$ and $t > 2$).

**10** The path traced out by a marked point on the rim of a wheel of radius $a$ when the wheel is rolled along a flat surface is called a cycloid.

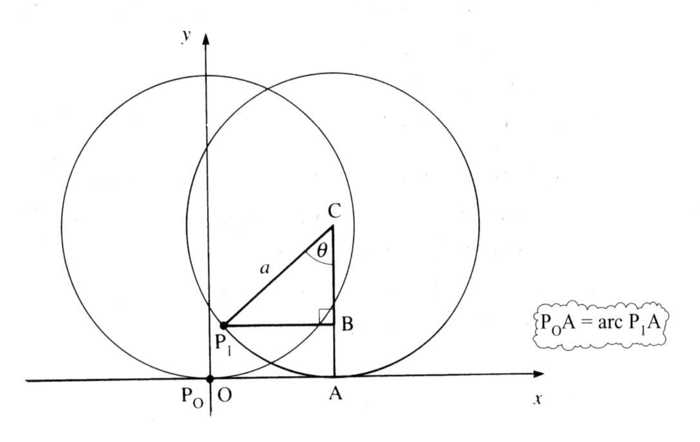

The diagram shows the wheel in its initial position when the lowest point on the rim is $P_O$, and when it has rotated through an angle $\theta$ (radians). In this position, the point $P_O$ has moved to $P_1$ with parametric equations given by

$$x = OA - P_1B = a\theta - a\sin\theta$$

$$y = AC - BC = a - a\cos\theta.$$

**(i)** Find the co-ordinates of the points corresponding to values of $\theta$ from 0 to $6\pi$ at intervals of $\frac{\pi}{3}$.

**(ii)** Sketch the curve for $0 \leqslant \theta \leqslant 6\pi$.

**(iii)** What do you notice about the curve?

**11** The curve with parametric equations

$$x = a\cos^3\theta \qquad y = a\sin^3\theta$$

is called an astroid.

**(i)** Sketch the curve.

**(ii)** On the same diagram sketch the curve

$$x = a\cos^n\theta \qquad y = a\sin^n\theta$$

for $n = 1, 2, 3, 4, 5, 6$. What happens if $n = 0$?

**(iii)** What can you say regarding the shape and position of the curve when $n \geqslant 7$ and **(a)** $n$ is even, **(b)** $n$ is odd?

## INVESTIGATIONS

### CUTTING OUT PATTERNS

A soft ball is to be made from felt, as in figure 4.14. The surface of the finished ball is composed of 16 equal sections, and is approximately spherical with a radius of 8 cm.

Investigate the shape needed for each of the sections and draw out a pattern that you can use to cut them out from flat pieces of felt.

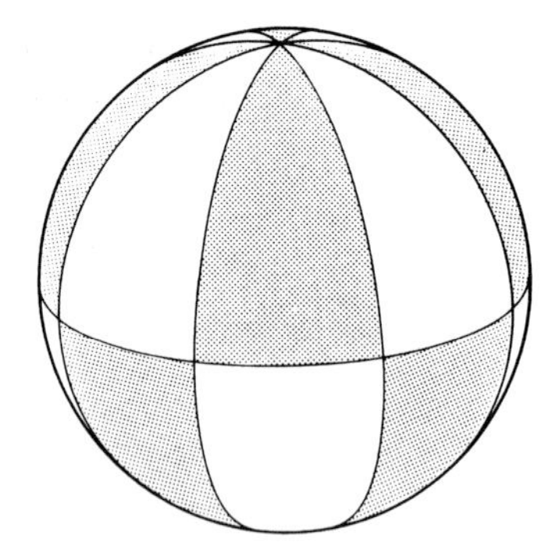

**Figure 4.14**

### THE APPARENT MOTION OF PLANETS

Most of the planets go round the Sun in elliptical (but nearly circular) orbits, and lie in very nearly the same plane. In this investigation you should assume the orbits are circular and you will find it helpful to work, at least to start with, with the suggested approximate data given in the table.

| Planet | Mean radius of orbit (km) | | Length of year: time for one rotation of the Sun (Earth days) | |
|---|---|---|---|---|
| | Accurate | Approximate | Accurate | Approximate |
| Mercury | $5.79 \times 10^7$ | $6 \times 10^7$ | 87.97 | 90 |
| Earth | $14.96 \times 10^7$ | $15 \times 10^7$ | 365.26 | 360 |
| Mars | $22.79 \times 10^7$ | $23 \times 10^7$ | 686.98 | 690 |

As seen from the Earth, it appears that the Sun is moving in a circle with the other planets circling around it.

**(i)** Find parametric equations for the paths of Mercury and Mars as seen from Earth, and so sketch their paths.

**(ii)** What is the effect of taking approximate values for the radius of orbit and length of year?

**(iii)** If you observe a planet at night time over a period of weeks or months you will see that it appears to move across the pattern of background stars. However, at times it will stop and move backwards (retrograde) before resuming its forward motion. How do your sketches of the planets' paths allow you to explain this phenomenon?

**(iv)** Some astronomy books will tell you that only the superior planets (those further from the Sun than Earth) are retrograde. Is this true, and if not how could such a mistake be made?

## Parametric differentiation

To differentiate a function which is defined in terms of a parameter $t$, you need to use the chain rule:

$$\frac{\mathrm{d}y}{\mathrm{d}x} = \frac{\mathrm{d}y}{\mathrm{d}t} \times \frac{\mathrm{d}t}{\mathrm{d}x}.$$

Since

$$\frac{\mathrm{d}t}{\mathrm{d}x} = \frac{1}{\frac{\mathrm{d}x}{\mathrm{d}t}}$$

it follows that

$$\frac{\mathrm{d}y}{\mathrm{d}x} = \frac{\frac{\mathrm{d}y}{\mathrm{d}t}}{\frac{\mathrm{d}x}{\mathrm{d}t}}$$

provided that $\frac{\mathrm{d}x}{\mathrm{d}t} \neq 0$.

**EXAMPLE 4.7**

A curve has the parametric equations $x = t^2$, $y = 2t$. Find

**(i)** $\dfrac{dy}{dx}$ in terms of the parameter $t$;

**(ii)** the equation of the tangent to the curve at the general point $(t^2, 2t)$;

**(iii)** the equation of the tangent at the point where $t = 3$.

**(iv)** Eliminate the parameter, and hence sketch the curve and the tangent at the point where $t = 3$.

**SOLUTION**

**(i)** $x = t^2 \quad \Rightarrow \quad \dfrac{dx}{dt} = 2t$

$y = 2t \quad \Rightarrow \quad \dfrac{dy}{dt} = 2$

$$\frac{dy}{dx} = \frac{\frac{dy}{dt}}{\frac{dx}{dt}} = \frac{2}{2t} = \frac{1}{t}.$$

The gradient of the curve at $(t^2, 2t)$

**(ii)** Using $y - y_1 = m(x - x_1)$ and taking the point $(x_1, y_1)$ as $(t^2, 2t)$, the equation of the tangent at the point $(t^2, 2t)$ is

$$y - 2t = \frac{1}{t}(x - t^2)$$

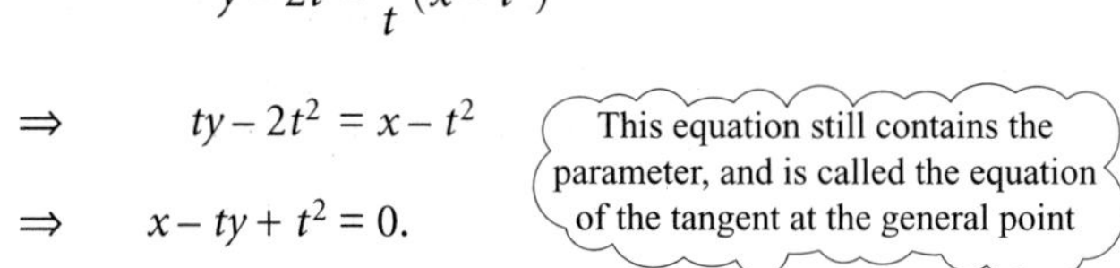

$$\Rightarrow \quad ty - 2t^2 = x - t^2$$

$$\Rightarrow \quad x - ty + t^2 = 0.$$



**(iii)** Substituting $t = 3$ into this equation gives the equation of the tangent at the point where $t = 3$.

The tangent is $x - 3y + 9 = 0$.

**(iv)** Eliminating $t$ from $x = t^2$, $y = 2t$ gives

$$x = \left(\frac{y}{2}\right)^2 \quad \text{or} \quad y^2 = 4x.$$

This a parabola whose line of symmetry is the $x$ axis.

The point where $t = 3$ has co-ordinates $(9, 6)$.

The tangent $x - 3y + 9 = 0$ crosses the axes at $(0, 3)$ and $(-9, 0)$.

The curve is shown in figure 4.15 (overleaf).

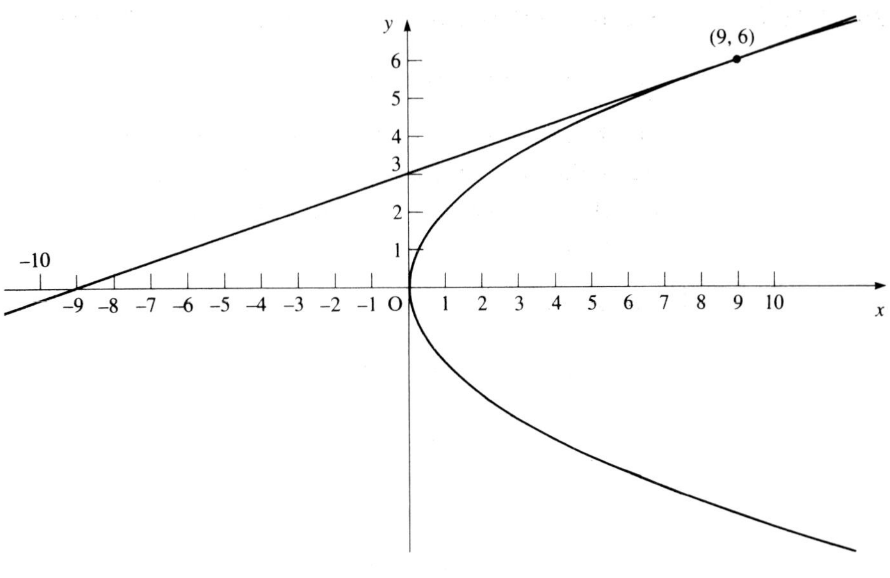


**Figure 4.15**



**EXAMPLE 4.8**

An ellipse has parametric equations $x = 4\cos\theta$, $y = 3\sin\theta$. Find

**(i)** $\dfrac{\mathrm{d}y}{\mathrm{d}x}$ at the point with parameter $\theta$;

**(ii)** the equation of the normal at the general point $(4\cos\theta, 3\sin\theta)$;

**(iii)** the equation of the normal at the point where $\theta = \frac{\pi}{4}$;

**(iv)** the co-ordinates of the point where $\theta = \frac{\pi}{4}$.

**(v)** Show the ellipse and the normal on a sketch.

**SOLUTION**

**(i)** $x = 4\cos\theta \quad \Rightarrow \quad \dfrac{\mathrm{d}x}{\mathrm{d}\theta} = -4\sin\theta$

$y = 3\sin\theta \quad \Rightarrow \quad \dfrac{\mathrm{d}y}{\mathrm{d}\theta} = 3\cos\theta$

$$\frac{\mathrm{d}y}{\mathrm{d}x} = \frac{\frac{\mathrm{d}y}{\mathrm{d}\theta}}{\frac{\mathrm{d}x}{\mathrm{d}\theta}} = \frac{3\cos\theta}{-4\sin\theta}$$

$$= -\frac{3\cos\theta}{4\sin\theta}.$$

**(ii)** The tangent and normal are perpendicular, so the gradient of the normal is

$$-\frac{1}{\frac{\mathrm{d}y}{\mathrm{d}x}} \quad \text{which is} \quad +\frac{4\sin\theta}{3\cos\theta}.$$

$m_1 m_2 = -1$ for perpendicular lines

**(ii)** Using $y - y_1 = m(x - x_1)$ and taking the point $(x_1, y_1)$ as $(4\cos\theta, 3\sin\theta)$, the equation of the normal at the point $(4\cos\theta, 3\sin\theta)$ is

$$y - 3\sin\theta = \frac{4\sin\theta}{3\cos\theta}(x - 4\cos\theta)$$

$$\Rightarrow \quad 3y\cos\theta - 9\sin\theta\cos\theta = 4x\sin\theta - 16\sin\theta\cos\theta$$

$$\Rightarrow \quad 4x\sin\theta - 3y\cos\theta - 7\sin\theta\cos\theta = 0.$$

**(iii)** When $\theta = \frac{\pi}{4}$, $\cos\theta = \frac{1}{\sqrt{2}}$ and $\sin\theta = \frac{1}{\sqrt{2}}$, so the equation of the normal is

$$4x \times \frac{1}{\sqrt{2}} - 3y \times \frac{1}{\sqrt{2}} - 7 \times \frac{1}{\sqrt{2}} \times \frac{1}{\sqrt{2}} = 0$$

$$\Rightarrow \quad 4\sqrt{2}x - 3\sqrt{2}y - 7 = 0$$

$$\Rightarrow \quad 4x - 3y - 4.95 = 0 \quad \text{(to 2 decimal places).}$$

**(iv)** The co-ordinates of the point where $\theta = \frac{\pi}{4}$ are

$$\left(4\cos\tfrac{\pi}{4}, 3\sin\tfrac{\pi}{4}\right) = \left(4 \times \tfrac{1}{\sqrt{2}}, 3 \times \tfrac{1}{\sqrt{2}}\right)$$

$$\approx (2.83, 2.12).$$

**(v)**

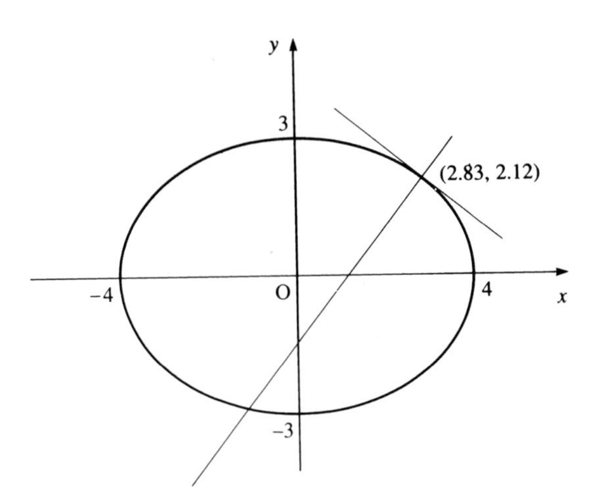


**Figure 4.16**

## Turning points

When the equation of a curve is given parametrically, the easiest way to distinguish between turning points is usually to consider the sign of $\frac{dy}{dx}$. If you use this method, you must be careful to ensure that you take points which are to the left and right of the turning point, i.e. have $x$ co-ordinates smaller and larger than those at the turning point. These will not necessarily be points whose parameters are smaller and larger than those at the turning point.

Alternatively, to find $\frac{d^2y}{dx^2}$ when $\frac{dy}{dx}$ is expressed in terms of a parameter requires a further use of the chain rule:

$$\frac{d^2y}{dx^2} = \frac{d}{dx}\left(\frac{dy}{dx}\right) = \frac{d}{dt}\left(\frac{dy}{dx}\right) \times \frac{dt}{dx}.$$

**EXAMPLE 4.9**

Find the turning points of the curve with parametric equations $x = 2t + 1$, $y = 3t - t^3$, and distinguish between them.

**SOLUTION**

$$x = 2t + 1 \quad \Rightarrow \quad \frac{dx}{dt} = 2$$

$$y = 3t - t^3 \quad \Rightarrow \quad \frac{dy}{dt} = 3 - 3t^2$$

$$\frac{dy}{dx} = \frac{\frac{dy}{dt}}{\frac{dx}{dt}} = \frac{3 - 3t^2}{2} = \frac{3(1 - t^2)}{2}.$$

Turning points occur when $\frac{dy}{dx} = 0$:

$$\Rightarrow \quad t^2 = 1$$

$$\Rightarrow \quad t = 1 \quad \text{or} \quad t = -1.$$

Now $\frac{d^2y}{dx^2} = \frac{d}{dx}\left(\frac{dy}{dx}\right) = \frac{d}{dt}\left(\frac{dy}{dx}\right) \times \frac{dt}{dx}$

so $\frac{d^2y}{dx^2} = \frac{d}{dt}\left[\frac{3(1 - t^2)}{2}\right] \times \frac{1}{2}$

$$= \tfrac{3}{2}(-2t) \times \tfrac{1}{2}$$

$$= -\tfrac{3}{2}t.$$

When $t = 1$: the point is (3, 2) and $\frac{d^2y}{dx^2} < 0$, so the turning point is a maximum.

When $t = -1$: the point is (–1, –2) and $\frac{d^2y}{dx^2} > 0$, so the turning point is a minimum.

**EXERCISE 4B**

**1** For each of the following curves, find $\frac{dy}{dx}$ in terms of the parameter.

**(i)** $x = 3t^2$
$y = 2t^3$

**(ii)** $x = \theta - \cos\theta$
$y = \theta + \sin\theta$

**(iii)** $x = t + \frac{1}{t}$
$y = t - \frac{1}{t}$

**(iv)** $x = 3\cos\theta$
$y = 2\sin\theta$

**(v)** $x = (t+1)^2$
$y = (t-1)^2$

**(vi)** $x = \theta\sin\theta + \cos\theta$
$y = \theta\cos\theta - \sin\theta$

**(vii)** $x = e^{2t} + 1$
$y = e^t$

**(viii)** $x = \frac{t}{1+t}$
$y = \frac{t}{1-t}$

**2** A curve has the parametric equations $x = \tan\theta$, $y = \tan 2\theta$. Find

**(i)** the value of $\frac{dy}{dx}$ when $\theta = \frac{\pi}{6}$;

**(ii)** the equation of the tangent to the curve at the point where $\theta = \frac{\pi}{6}$;

**(iii)** the equation of the normal to the curve at the point where $\theta = \frac{\pi}{6}$.

**3** A curve has the parametric equations $x = t^2$, $y = 1 - \frac{1}{2t}$ for $t > 0$. Find

**(i)** the co-ordinates of the point P where the curve cuts the $x$ axis;

**(ii)** the gradient of the curve at this point;

**(iii)** the equation of the tangent to the curve at P;

**(iv)** the co-ordinates of the point where the tangent cuts the $y$ axis.

**4** A curve has parametric equations $x = at^2$, $y = 2at$, where $a$ is constant. Find

**(i)** the equation of the tangent to the curve at the point with parameter $t$;

**(ii)** the equation of the normal to the curve at the point with parameter $t$;

**(iii)** the co-ordinates of the points where the normal cuts the $x$ and $y$ axes.

**5** A curve has parametric equations $x = \cos\theta$, $y = \cos 2\theta$.

**(i)** Show that $\frac{dy}{dx} = 4\cos\theta$.

**(ii)** By writing $\frac{dy}{dx}$ in terms of $x$, show that $\frac{d^2y}{dx^2} - 4 = 0$.

**6** The parametric equations of a curve are $x = at$, $y = \frac{b}{t}$, where $a$ and $b$ are constant. Find in terms of $a$, $b$ and $t$

**(i)** $\frac{dy}{dx}$;

**(ii)** the equation of the tangent to the curve at the general point $(at, \frac{b}{t})$;

**(iii)** the co-ordinates of the points X and Y where the tangent cuts the $x$ and $y$ axes.

**(iv)** Show that the area of triangle OXY is constant, where O is the origin.

**7** The equation of a curve is given in terms of the parameter $t$ by the equations $x = 4t$ and $y = 2t^2$ where $t$ takes positive and negative values.

**(i)** Sketch the curve.

P is the point on the curve with parameter $t$.

**(ii)** Show that the gradient at P is $t$.

**(iii)** Find and simplify the equation of the tangent at P.

The tangents at two points Q (with parameter $t_1$) and R (with parameter $t_2$) meet at S.

**(iv)** Find the co-ordinates of S.

**(v)** In the case when $t_1 + t_2 = 2$ show that S lies on a straight line. Give the equation of the line.

[MEI]

**8** The diagram shows a sketch of the curve given parametrically in terms of $t$ by the equations $x = 1 - t^2$, $y = 2t + 1$.

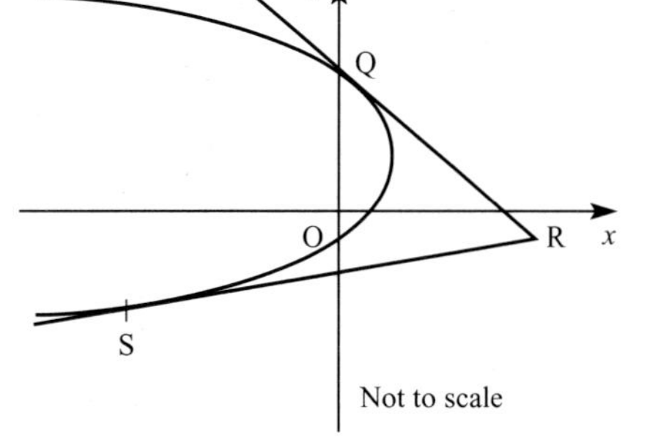


**(i)** Show that the point Q (0, 3) lies on the curve, stating the value of $t$ corresponding to this point.

**(ii)** Show that, at the point with parameter $t$,

$$\frac{dy}{dx} = -\frac{1}{t}.$$

**(iii)** Find the equation of the tangent at Q.

**(iv)** Verify that the tangent at Q passes through the point R (4, –1).

**(v)** The other tangent from R to the curve touches the curve at the point S and has equation $3y - x + 7 = 0$. Find the co-ordinates of S.

[MEI]

**9** The diagram shows a sketch of the curve with parametic equations $x = 1 - 2t$, $y = t^2$. The tangent and normal at P are also shown.

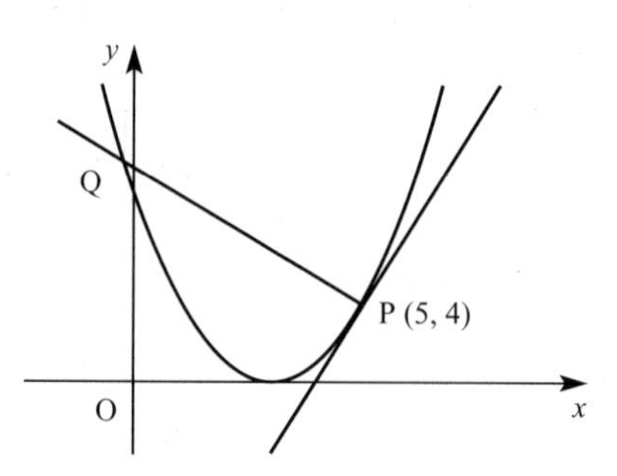


**(i)** Show that the point P (5, 4) lies on the curve by stating the value of $t$ corresponding to this point.

**(ii)** Show that, at the point with parameter $t$, $\frac{dy}{dx} = -t$.

**(iii)** Find the equation of the tangent at P.

**(iv)** The normal at P cuts the curve again at Q. Find the co-ordinates of Q.

[MEI]

**10** A point P moves in a plane so that at time $t$ its co-ordinates are given by $x = 4\cos t$, $y = 3\sin t$. Find

**(i)** $\frac{dy}{dx}$ in terms of $t$;

**(ii)** the equation of the tangent to its path at time $t$;

**(iii)** the values of $t$ for which the particle is travelling parallel to the line $x + y = 0$.

**11** A circle has parametric equations $x = 3 + 2\cos\theta$, $y = 3 + 2\sin\theta$.

**(i)** Find the equation of the tangent at the point with parameter $\theta$.

**(ii)** Show that this tangent will pass through the origin provided that $\sin\theta + \cos\theta = -\frac{2}{3}$.

**(iii)** By writing $\sin\theta + \cos\theta$ in the form $R\sin(\theta + \alpha)$, solve the equation $\sin\theta + \cos\theta = -\frac{2}{3}$ for $0 \leqslant \theta \leqslant 2\pi$.

**(iv)** Illustrate the circle and tangents on a sketch, showing clearly the values of $\theta$ which you found in part (iii).

**12** The parametric equations of the circle centre (2, 5) and radius 3 units are $x = 2 + 3\cos\theta$, $y = 5 + 3\sin\theta$.

**(i)** Find the gradient of the circle at the point with parameter $\theta$.

**(ii)** Find the equation of the normal to the circle at this point.

**(iii)** Show that the normal at any point on the circle passes through the centre. (This is an alternative proof of the result 'tangent and radius are perpendicular'.)

**13** The parametric equations of a curve are

$$x = 3\cos\theta, \qquad y = 2\sin\theta \qquad \text{for } 0 \leqslant \theta < 2\pi.$$

**(i)** By eliminating $\theta$ between these two equations, find the cartesian equation of the curve.

**(ii)** Draw a sketch of the curve, giving the co-ordinates of the points where it cuts the axes. On your sketch show the pair of tangents which pass through the point (6, 2).

**(iii)** Use the parametric equations to calculate $\frac{dy}{dx}$ in terms of $\theta$.

You are given that the equation of the tangent to the curve at $(3\cos\theta, 2\sin\theta)$ is

$$2x\cos\theta + 3y\sin\theta = 6.$$

**(iv)** Show that, for tangents to the curve which pass through the point (6, 2),

$$2\cos\theta + \sin\theta = 1.$$

**(v)** Solve the equation in (iv) to find the two values of $\theta$ (in radians correct to 2 decimal places) corresponding to the two tangents.

[MEI]

**14** An ellipse has equation given in parametric form by $x = 4\cos\phi$, $y = 3\sin\phi$, $0 \leqslant \phi < 2\pi$.

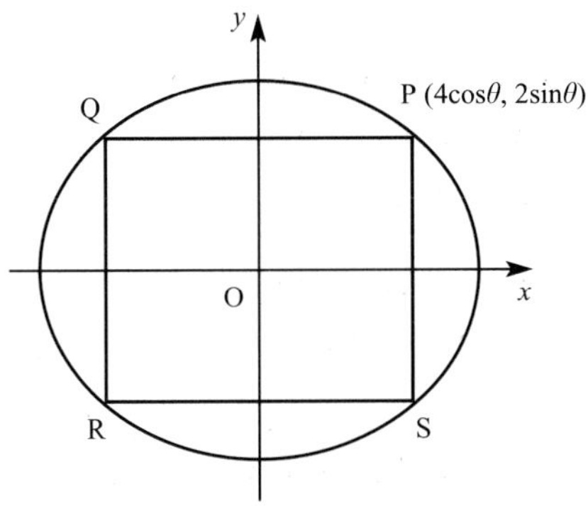


The sketch illustrates this ellipse and point P $(4\cos\theta, 3\sin\theta)$, $0 \leqslant \theta \leqslant \frac{\pi}{2}$. Rectangle PQRS has PQ parallel to the $x$ axis and PS parallel to the $y$ axis, with Q, R and S also on the ellipse.

**(i) (a)** Express the equation of the ellipse in cartesian form.

**(b)** The length, $L$, of the perimeter of PQRS is given by $L = 12\sin\theta + 16\cos\theta$. Express $L$ in the form $r\sin(\theta + \alpha)$, where $r$ and $\alpha$ are constants to be determined.

**(c)** Find the maximum value of $L$ and the value of $\theta$, $0 \leqslant \theta \leqslant \frac{\pi}{2}$, for which it occurs.

**(ii)** The line PS produced meets the line $y = -8$ at the point U with co-ordinates $(4\cos\theta, -8)$, where $0 \leqslant \theta \leqslant \frac{\pi}{2}$.

**(a)** Write down the gradient of OU.

**(b)** Calculate the gradient of the tangent at P.

**(c)** Find the value of $\theta$ for which OU is parallel to the tangent at P. Give your answer correct to 2 decimal places.

[MEI]

**15** The curve shown in the diagram has parametric equations

$$x = \frac{1}{1+t}, \qquad y = \frac{1}{(1+t)(1-t)}, \qquad (t \neq \pm 1).$$

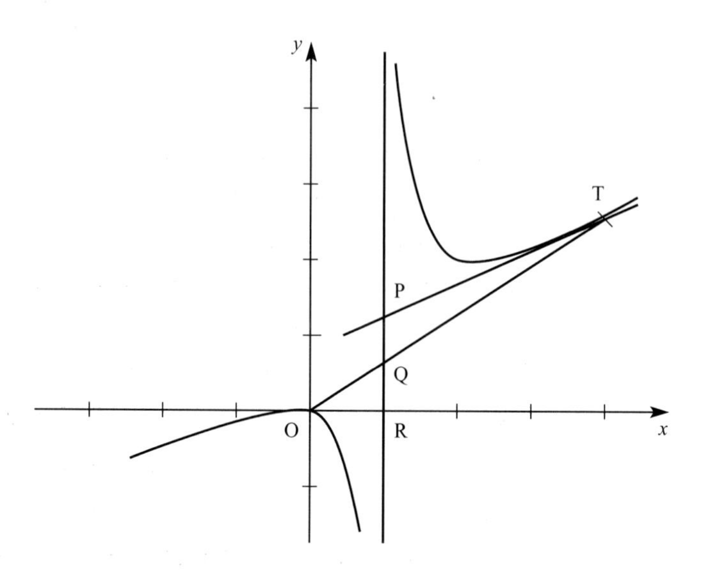


**(i)** Express $t$ in terms of $x$. Hence show that the cartesian equation of the curve is $y = \dfrac{x^2}{2x-1}$.

**(ii)** Find $\dfrac{dy}{dt}$ and $\dfrac{dx}{dt}$ and hence show that

$$\frac{dy}{dx} = -\frac{2t}{(1-t)^2}.$$

**(iii)** You are given that the equation of the tangent at the point T having parameter $t$ is $2tx + (1-t)^2 y = 1$.

Find the $y$ co-ordinate of the point P where this tangent cuts the line $x = \frac{1}{2}$.

**(iv)** The point Q is the intersection of the line $x = \frac{1}{2}$ and the straight line joining the origin to the point T. Point R has co-ordinates $(\frac{1}{2}, 0)$.

Show that RQ = QP.

[MEI]

**16**

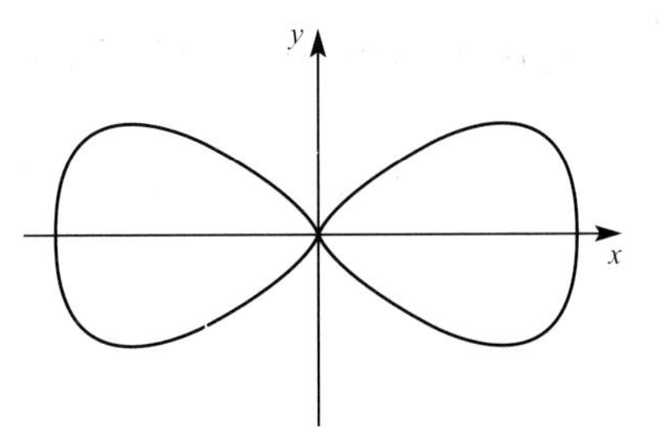


The diagram shows the curve with parametric equations

$$x = \cos t, \qquad y = \tfrac{1}{2}\sin 2t$$

for $0 \leqslant t \leqslant 2\pi$. The curve is symmetrical about both axes.

**(i)** Copy the diagram. Locate and label on your sketch the points having parameters

$$t = 0, \quad t = \tfrac{\pi}{2}, \quad t = \pi \text{ and } t = \tfrac{3\pi}{2}.$$

**(ii)** Find an expression for $\dfrac{dy}{dx}$ in terms of the parameter $t$. Hence show that, at the origin, the curve crosses itself at right angles.

**(iii)** Show that the cartesian equation of the curve is $y^2 = x^2(1 - x^2)$.

**(iv)** Show that the parameters of the points where the gradient of the curve is $-\tfrac{7}{2}$ satisfy the equation $4\sin^2 t + 7\sin t - 2 = 0$.

Find the parameters of these points.

[MEI]

**17** The curve in the diagram is given by the parametric equations

$$x = 2\cos\theta + \sin\theta, \qquad y = \cos\theta + 2\sin\theta \qquad \text{for } 0 \leqslant \theta < 2\pi.$$

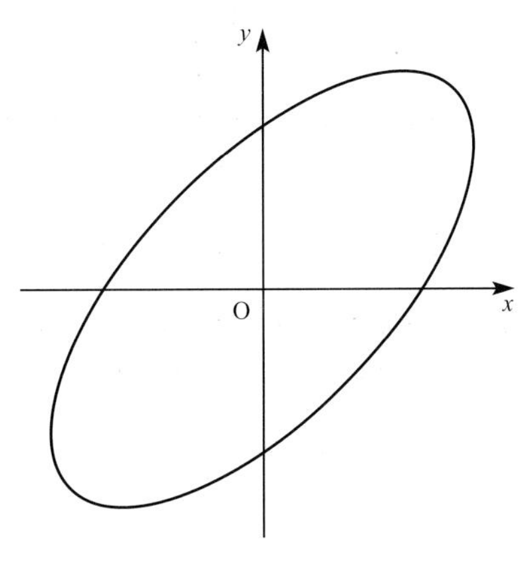


**(i)** Express $x$ in the form $R\cos(\theta - \alpha)$, where $R > 0$. Between what values must $x$ lie?

**(ii)** Find the gradient of the curve at the point where $\theta = \tfrac{\pi}{2}$.

**(iii)** Show that, for any point on the curve,

$$x^2 + y^2 = 5 + 4\sin 2\theta.$$

**(iv)** Find the greatest and least distances of a point on the curve from the origin.

[MEI]

**KEY POINTS**

**1** In parametric equations the relationship between two variables is expressed by writing both of them in terms of a third variable or *parameter*.

**2** To draw a graph from parametric equations, plot the points on the curve given by different values of the parameter.

**3** Eliminating the parameter gives the cartesian equation of the curve.

**4** The parametric equations of some well-known curves:

- **Circle** centre (0, 0) and radius $r$

  $$x = r\cos\theta \qquad y = r\sin\theta.$$

- **Circle** centre $(a, b)$ and radius $r$

  $$x = a + r\cos\theta \qquad y = b + r\sin\theta.$$

- **Ellipse** centre (0, 0) with major axis $2a$ and minor axis $2b$

  $$x = a\cos\theta \qquad y = b\sin\theta.$$

- **Parabola** whose line of symmetry is the $x$ axis

  $$x = at^2 \qquad y = 2at.$$

- **Rectangular hyperbola** $xy = c^2$

  $$x = ct \qquad y = \frac{c}{t}.$$

**5** $\dfrac{dy}{dx} = \dfrac{\frac{dy}{dt}}{\frac{dx}{dt}}$ provided that $\dfrac{dx}{dt} \neq 0$.

**6** $\dfrac{d^2y}{dx^2} = \dfrac{d}{dx}\left(\dfrac{dy}{dx}\right) = \dfrac{d}{dt}\left(\dfrac{dy}{dx}\right) \times \dfrac{dt}{dx}.$

# 5 Vector geometry

**We drove into the future looking into a rear view mirror.**

*Herbert Marshall McLuhan*

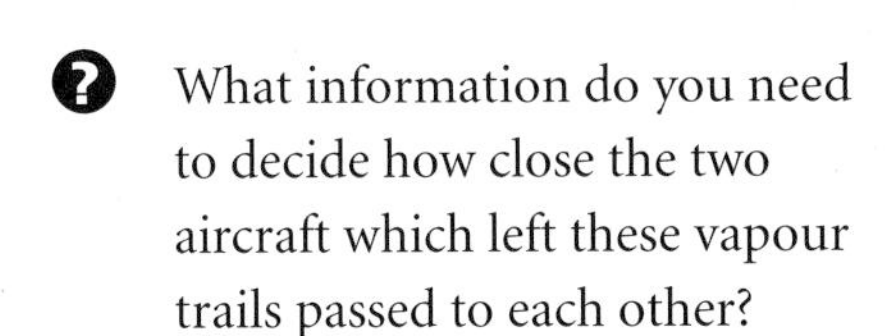

What information do you need to decide how close the two aircraft which left these vapour trails passed to each other?

## Vectors

A quantity which has both size and direction is called a *vector*. The velocity of an aircraft through the sky is an example of a vector, having size (e.g. 600 mph) and direction (on a course of 254°). By contrast the mass of the aircraft (100 tonnes) is completely described by its size and no direction is associated with it; such a quantity is called a *scalar*.

Vectors are used extensively in mechanics to represent quantities such as force, velocity and momentum, and in geometry to represent displacements. They are an essential tool in three-dimensional co-ordinate geometry and it is this application of vectors which is the subject of this chapter. However, before coming on to this, you need to be familiar with the associated vocabulary and notation, in two and three dimensions.

### Terminology

In two dimensions, it is common to represent a vector by a drawing of a straight line with an arrowhead. The length represents the size, or magnitude, of the vector and the direction is indicated by the line and the arrowhead. Direction is usually given as the angle the vector makes with the positive $x$ axis, with the anticlockwise direction taken to be positive.

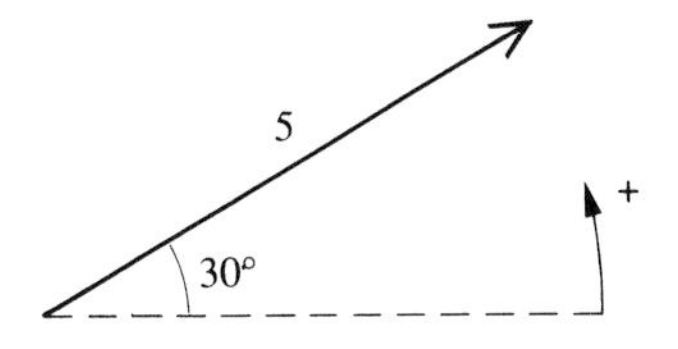


**Figure 5.1**

The vector in figure 5.1 has magnitude 5, direction +30°. This is written (5, 30°) and said to be in *magnitude–direction form* or in *polar form*. The general form of a vector written in this way is $(r, \theta)$ where $r$ is its magnitude and $\theta$ its direction.

*Note*

In the special case when the vector is representing real travel, as in the case of the velocity of an aircraft, the direction may be described by a compass bearing with the angle measured from north, clockwise. However, this is not done in this chapter, where directions are all taken to be measured anticlockwise from the positive $x$ direction.

An alternative way of describing a vector is in terms of *components* in given directions. The vector in figure 5.2 is 4 units in the $x$ direction, and 2 in the $y$ direction, and this is denoted by $\begin{pmatrix} 4 \\ 2 \end{pmatrix}$.

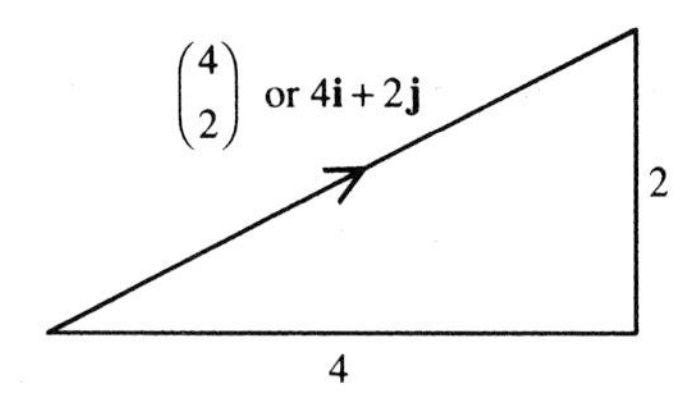


**Figure 5.2**

This may also be written as $4\mathbf{i} + 2\mathbf{j}$, where $\mathbf{i}$ is a vector of magnitude 1, a *unit vector*, in the $x$ direction and $\mathbf{j}$ is a unit vector in the $y$ direction (figure 5.3).

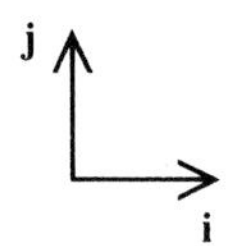


**Figure 5.3**

In a book, a vector may be printed in bold, for example **p** or **OP**, or as a line between two points with an arrow above it to indicate its direction, such as $\overrightarrow{OP}$. When you write a vector by hand, it is usual to underline it, for example, $\underline{p}$ or $\underline{OP}$, or to put an arrow above it, as in $\overrightarrow{OP}$.

To convert a vector from component form to magnitude–direction form, or vice versa, is just a matter of applying trigonometry to a right-angled triangle.

**EXAMPLE 5.1**

Write the vector $\mathbf{a} = 4\mathbf{i} + 2\mathbf{j}$ in magnitude–direction form.

**SOLUTION**

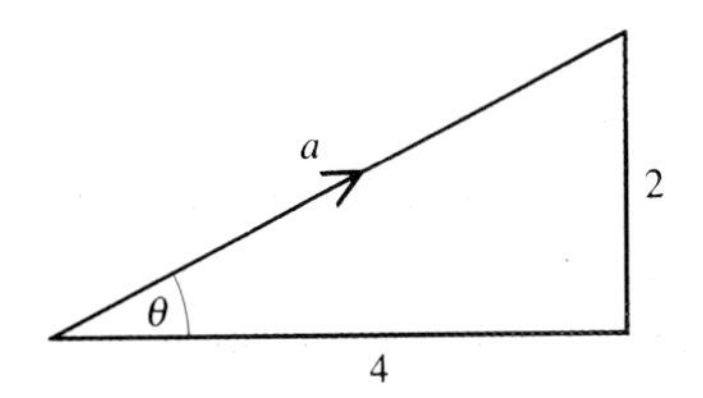


**Figure 5.4**

The magnitude of **a** is given by the length $a$ in figure 5.4:

$$a = \sqrt{4^2 + 2^2} \qquad \text{(using Pythagoras' theorem)}$$

$$= 4.47 \qquad \text{(to 3 significant figures).}$$

The direction is given by the angle $\theta$:

$$\tan\theta = \frac{2}{4} = 0.5$$

$$\theta = 26.6° \qquad \text{(to 3 significant figures).}$$

The vector **a** is (4.47, 26.6°).

The magnitude of a vector is also called its *modulus* and denoted by the symbols | |. In the example $\mathbf{a} = 4\mathbf{i} + 2\mathbf{j}$, the modulus of **a**, written | **a** |, is 4.47. Another convention for writing the magnitude of a vector is to use the same letter, but in italics and not bold type; thus the magnitude of **a** may be written $a$.

**EXAMPLE 5.2**

Write the vector (5, 60°) in component form.

**SOLUTION**

In the right-angled triangle OPX

$$\text{OX} = 5\cos 60° = 2.5$$

$$\text{XP} = 5\sin 60° = 4.33 \quad \text{(to 2 decimal places)}$$

$$\overrightarrow{\text{OP}} \text{ is } \begin{pmatrix} 2.5 \\ 4.33 \end{pmatrix} \quad \text{or} \quad 2.5\mathbf{i} + 4.33\mathbf{j}.$$

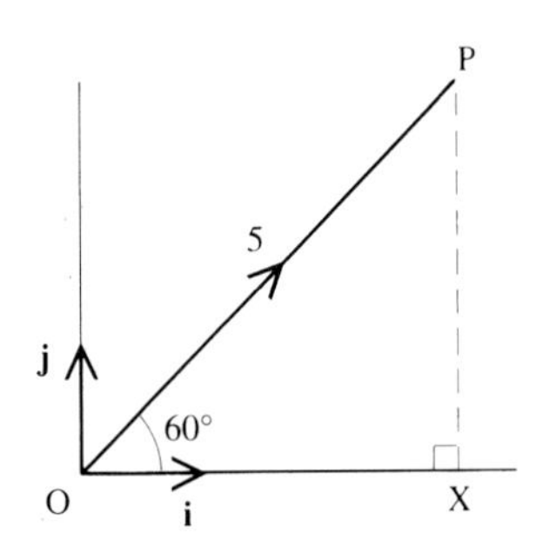


**Figure 5.5**

This technique can be written as a general rule, for all values of $\theta$:

$$(r, \theta) \rightarrow \begin{pmatrix} r\cos\theta \\ r\sin\theta \end{pmatrix} = (r\cos\theta)\mathbf{i} + (r\sin\theta)\mathbf{j}.$$

**EXAMPLE 5.3**

Write the vector (10, 290°) in component form.

**SOLUTION**

In this case $r = 10$ and $\theta = 290°$.

$$(10, 290°) \rightarrow \begin{pmatrix} 10\cos290° \\ 10\sin290° \end{pmatrix} = \begin{pmatrix} 3.42 \\ -9.40 \end{pmatrix} \quad \text{to 2 decimal places.}$$

This may also be written $3.42\mathbf{i} - 9.40\mathbf{j}$.

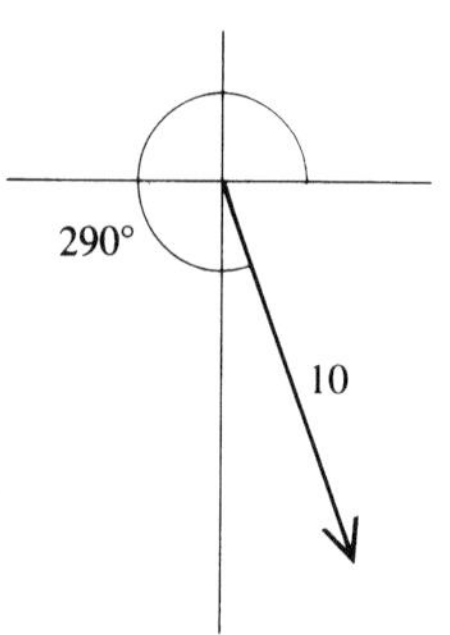


**Figure 5.6**

In Example 5.3 the signs looked after themselves. The component in the **i** direction came out positive, that in the **j** direction negative, as must be the case for a direction in the fourth quadrant ($270° < \theta < 360°$). This will always be the case when the conversion is from magnitude–direction form into component form.

The situation is not quite so straightforward when the conversion is carried out the other way, from component form to magnitude–direction form. In that case, it is best to draw a diagram and use it to see the approximate size of the angle required. This is shown in the next example.

**EXAMPLE 5.4**

Write $-5\mathbf{i} + 4\mathbf{j}$ in magnitude–direction form.

**SOLUTION**

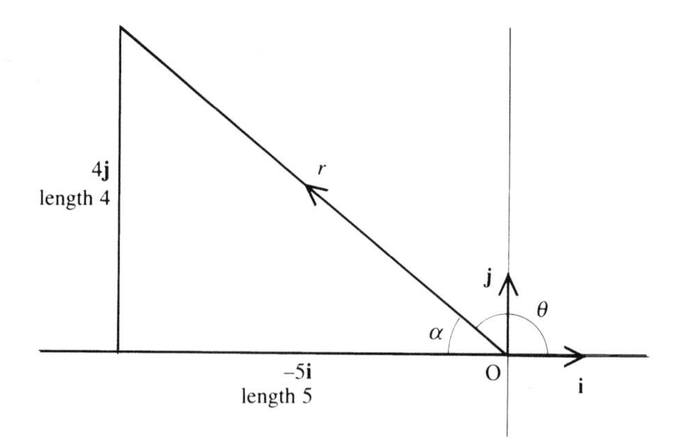


**Figure 5.7**

In this case, the magnitude $r = \sqrt{5^2 + 4^2} = \sqrt{41}$

$= 6.40$ (to 2 decimal places).

The direction is given by the angle $\theta$ in figure 5.7, but first find the angle $\alpha$:

$$\tan\alpha = \frac{4}{5} \quad \Rightarrow \quad \alpha = 38.7° \quad \text{(to nearest 0.1°)}$$

so $\theta = 180 - \alpha = 141.3°$.

The vector is (6.40, 141.3°) in magnitude–direction form.

## USING YOUR CALCULATOR

Most graphics calculators include the facility to convert from polar co-ordinates $(r, \theta)$ to rectangular co-ordinates $(x, y)$, and vice versa. This is the same as converting one form of a vector into the other. Once you are clear what is involved, you will probably prefer to do such conversions on your calculator.



## Equal vectors

The statement that two vectors **a** and **b** are equal means two things:

- The direction of **a** is the same as the direction of **b**.
- The magnitude of **a** is the same as the magnitude of **b**.

If the vectors are given in component form, each component of **a** equals the corresponding component of **b**.

## Position vectors

Saying the vector **a** is given by $4\mathbf{i} + 2\mathbf{j}$ tells you the components of the vector, or equivalently its magnitude and direction. It does not tell you where the vector is situated; indeed it could be anywhere.

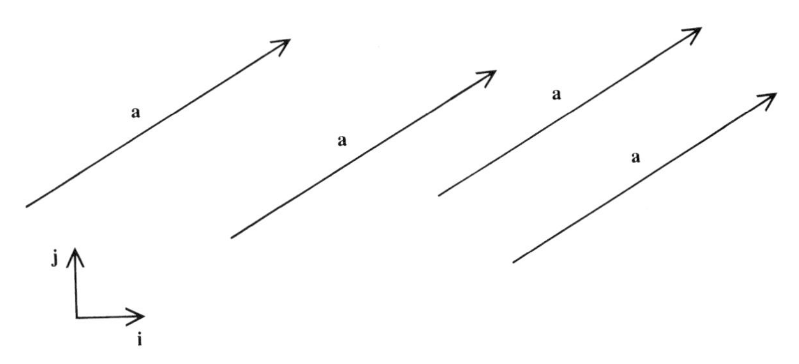


**Figure 5.8**

All of the lines in figure 5.8 represent the vector **a**.

There is, however, one special case which is an exception to the rule, that of a vector which starts at the origin. This is called a *position vector*. Thus the line joining the origin to the point (3, 5) is the position vector $\begin{pmatrix} 3 \\ 5 \end{pmatrix}$ or $3\mathbf{i} + 5\mathbf{j}$. Another way of expressing this is to say that the point (3, 5) has the position vector $\begin{pmatrix} 3 \\ 5 \end{pmatrix}$.

**EXAMPLE 5.5**

Points L, M and N have co-ordinates (4, 3), (–2, –1) and (2, 2).

**(i)** Write down, in component form, the position vector of L and the vector $\overrightarrow{MN}$.

**(ii)** What do your answers to (i) tell you about the lines OL and MN?

**SOLUTION**

**(i)** The position vector of L is $\overrightarrow{OL} = \begin{pmatrix} 4 \\ 3 \end{pmatrix}$.

The vector $\overrightarrow{MN}$ is also $\begin{pmatrix} 4 \\ 3 \end{pmatrix}$ (see figure 5.9).

**(ii)** Since $\overrightarrow{OL} = \overrightarrow{MN}$, lines OL and MN are parallel and equal in length.

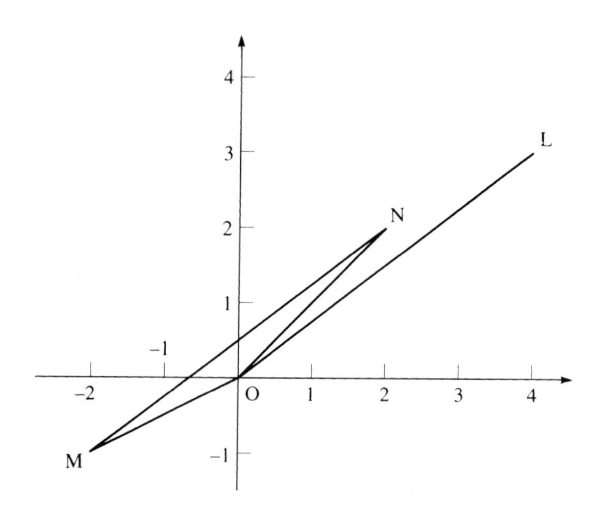


**Figure 5.9**

*Note*

A line joining two points, like MN in figure 5.9, is often called a *line segment*, meaning that it is just that particular part of the infinite straight line that passes through those two points.

## EXERCISE 5A

**1** Express the following vectors in component form.

**(i)**

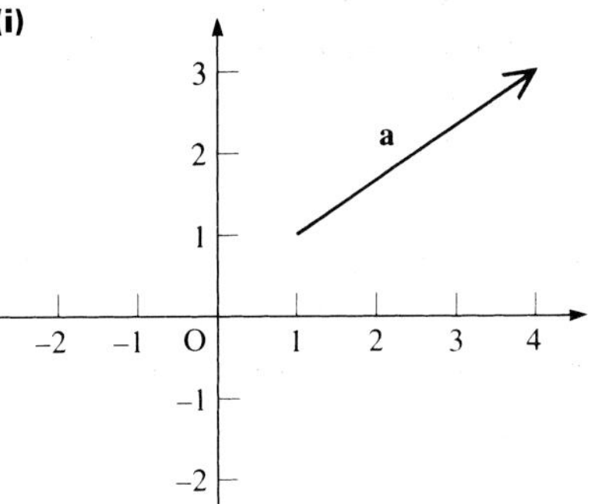


**(ii)**

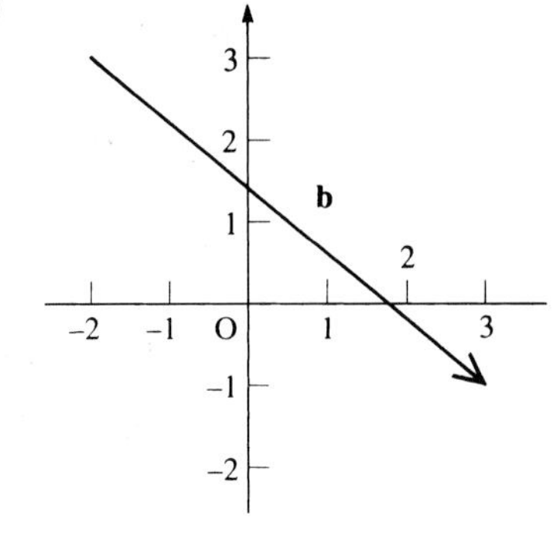


**(iii)**

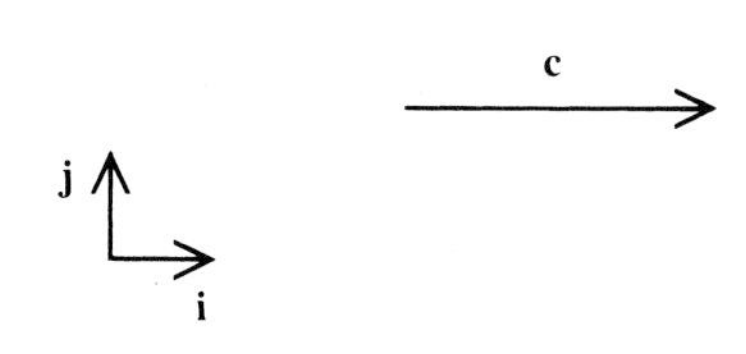


**(iv)**

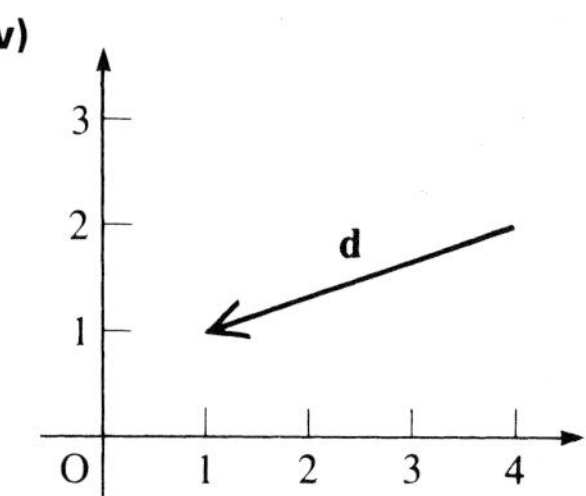


**(v)**

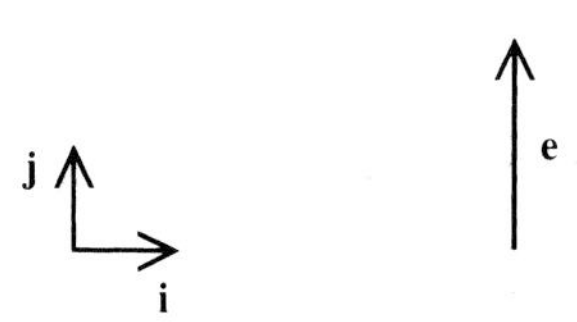


**2** Draw diagrams to show these vectors and then write them in magnitude–direction form. You may find it helpful to use your calculator to check your answers.

**(i)** $2\mathbf{i} + 3\mathbf{j}$ **(ii)** $\begin{pmatrix} 3 \\ -2 \end{pmatrix}$ **(iii)** $\begin{pmatrix} -4 \\ -4 \end{pmatrix}$

**(iv)** $-\mathbf{i} + 2\mathbf{j}$ **(v)** $3\mathbf{i} - 4\mathbf{j}$

**3** Draw diagrams to show these vectors and then write them in component form. You may find it helpful to use your calculator to check your answers.

**(i)** $(5, 45°)$ **(ii)** $(10, 210°)$

**(iii)** $(4, \frac{\pi}{2})$ **(iv)** $(8, 2\pi)$

**(v)** $(4, \frac{5\pi}{4})$

**4** Write, in component form, the vectors represented by the line segments joining

**(i)** (2, 3) to (4, 1)

**(ii)** (4, 0) to (6, 0)

**(iii)** (0, 0) to (0, –4)

**(iv)** (0, –4) to (0, 0)

**(v)** (–3, –4) to (–4, –3)

**(vi)** (–4, –3) to (–3, –4)

**(vii)** (0, 0) to (8, 0)

**(viii)** (8, 0) to (0, 0)

**(ix)** (3, 1) to (5, –3)

**(x)** (3, –1) to (7, 3)

**5** The points A, B and C have co-ordinates (2, 3), (0, 4) and (–2, 1).

**(i)** Write down the position vectors of A and C.

**(ii)** Write down the vectors of the line segments joining AB and CB.

**(iii)** What do your answers to (i) and (ii) tell you about

**(a)** AB and OC?

**(b)** CB and OA?

**(iv)** Describe the quadrilateral OABC.

## Multiplying a vector by a scalar

When a vector is multiplied by a number (a scalar) its length is altered but its direction remains the same.

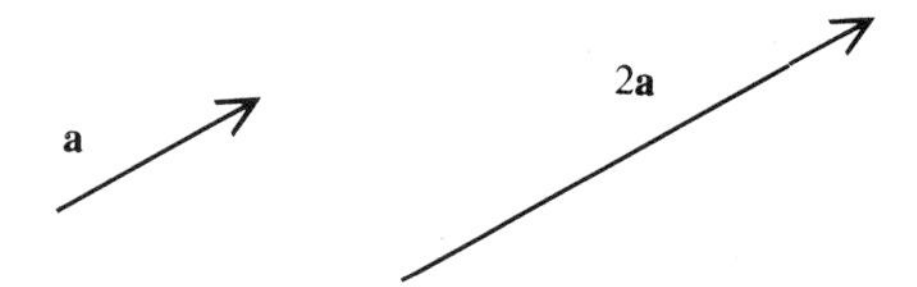


**Figure 5.10**

The vector 2**a** in figure 5.10 is twice as long as the vector **a** but in the same direction.

When the vector is in component form, each component is multiplied by the number. For example:

$$2 \times (3\mathbf{i} - 5\mathbf{j}) = 6\mathbf{i} - 10\mathbf{j}$$

$$2 \times \begin{pmatrix} 3 \\ -5 \end{pmatrix} = \begin{pmatrix} 6 \\ -10 \end{pmatrix}.$$

### The negative of a vector

In figure 5.11 the vector $-\mathbf{a}$ has the same length as the vector $\mathbf{a}$ but the opposite direction.

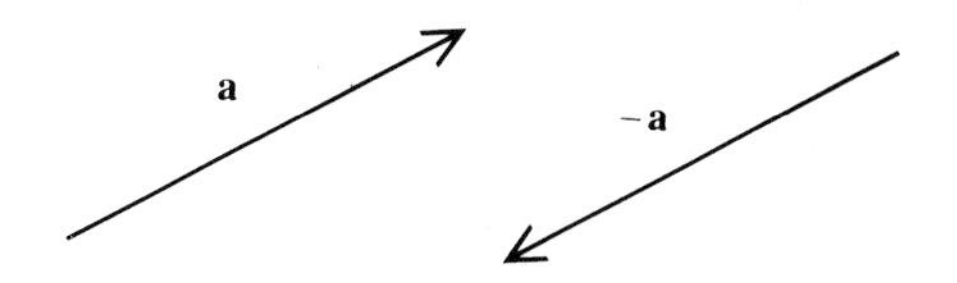


**Figure 5.11**

When $\mathbf{a}$ is given in component form, the components of $-\mathbf{a}$ are the same as those for $\mathbf{a}$ but with their signs reversed. So

$$-\begin{pmatrix} 23 \\ -11 \end{pmatrix} = \begin{pmatrix} -23 \\ +11 \end{pmatrix}.$$

### Adding vectors

When vectors are given in component form, they can be added component by component. This process can be seen geometrically by drawing them on graph paper, as in the example below.

**EXAMPLE 5.6**

Add the vectors $2\mathbf{i} - 3\mathbf{j}$ and $3\mathbf{i} + 5\mathbf{j}$.

**SOLUTION**

$2\mathbf{i} - 3\mathbf{j} + 3\mathbf{i} + 5\mathbf{j} = 5\mathbf{i} + 2\mathbf{j}$

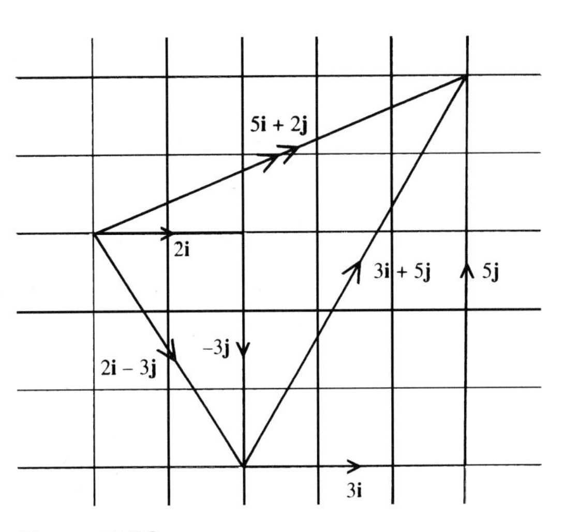


**Figure 5.12**

The sum of two (or more) vectors is called the *resultant* and is usually indicated by being marked with two arrowheads.

Adding vectors is like adding the legs of a journey to find its overall outcome (figure 5.13).

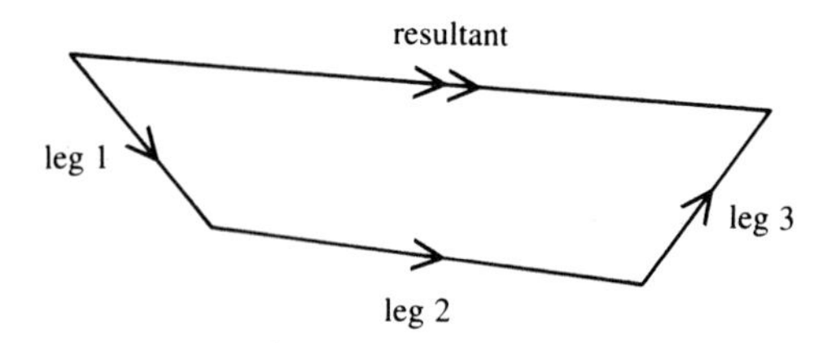


**Figure 5.13**

When vectors are given in magnitude–direction form, you can find their resultant by making a scale drawing, as in figure 5.13. If, however, you need to calculate their resultant, it is usually easiest to convert the vectors into component form, add component by component, and then convert the answer back to magnitude–direction form.

## Subtracting vectors

Subtracting one vector from another is the same as adding the negative of the vector.

**EXAMPLE 5.7**

Two vectors **a** and **b** are given by

$$\mathbf{a} = 2\mathbf{i} + 3\mathbf{j} \qquad \mathbf{b} = -\mathbf{i} + 2\mathbf{j}.$$

**(i)** Find **a** – **b**.

**(ii)** Draw diagrams showing **a**, **b**, **a** – **b**.

**SOLUTION**

**(i)** $\mathbf{a} - \mathbf{b} = (2\mathbf{i} + 3\mathbf{j}) - (-\mathbf{i} + 2\mathbf{j})$
$= 3\mathbf{i} + \mathbf{j}$

**(ii)**

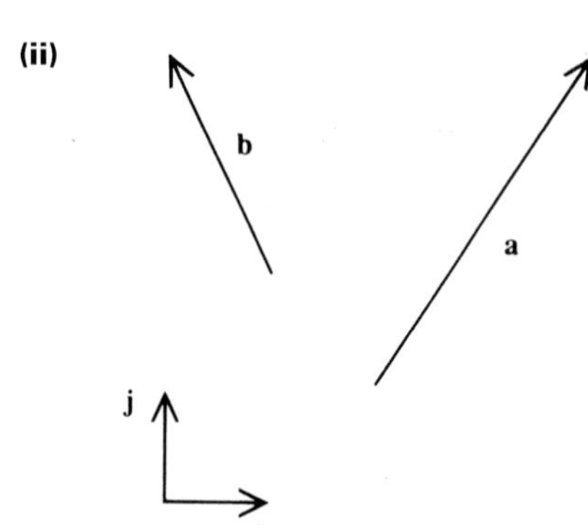


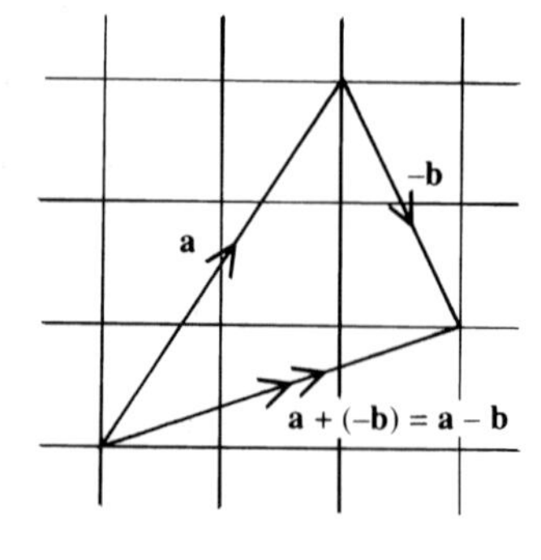


**Figure 5.14**

When you find the vector represented by the line segment joining two points, you are in effect subtracting their position vectors. If, for example,

P is the point (2, 1) and Q is the point (3, 5), $\overrightarrow{PQ}$ is $\begin{pmatrix}1\\4\end{pmatrix}$, as figure 5.15 shows.

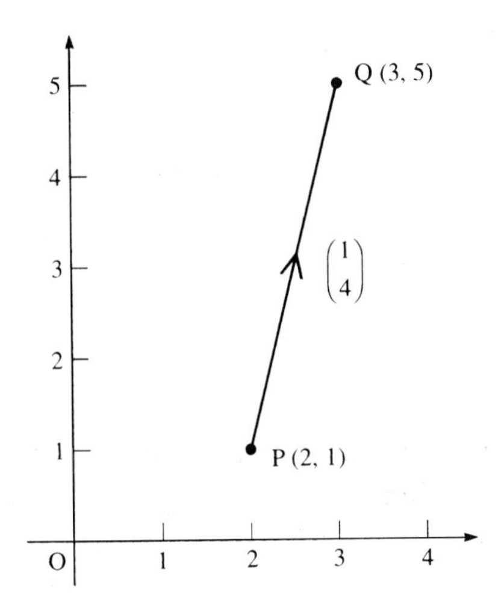


**Figure 5.15**

You find this by saying

$$\overrightarrow{PQ} = \overrightarrow{PO} + \overrightarrow{OQ} = -\mathbf{p} + \mathbf{q}.$$

In this case, this gives

$$\overrightarrow{PQ} = -\begin{pmatrix}2\\1\end{pmatrix} + \begin{pmatrix}3\\5\end{pmatrix} = \begin{pmatrix}1\\4\end{pmatrix}$$

as expected.

This is an important result, that

$$\overrightarrow{PQ} = \mathbf{q} - \mathbf{p}$$

where **p** and **q** are the position vectors of P and Q.

## Geometrical figures

It is often useful to be able to express lines in a geometrical figure in terms of given vectors, as in the next example.

**EXAMPLE 5.8**

Figure 5.16 shows a hexagon ABCDEF.

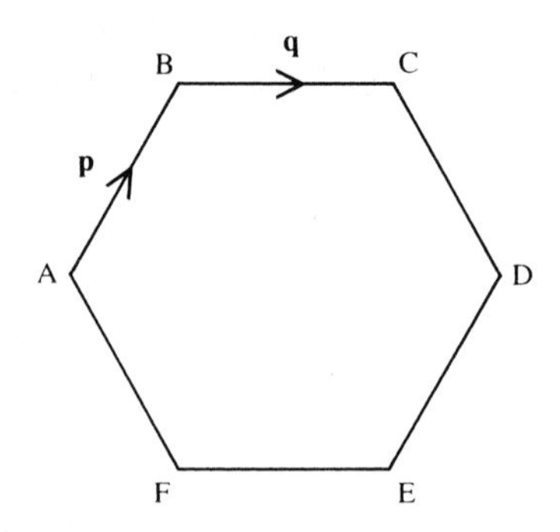


**Figure 5.16**

The hexagon is regular and consequently $\overrightarrow{AD} = 2\overrightarrow{BC}$.
$\overrightarrow{AB} = \mathbf{p}$ and $\overrightarrow{BC} = \mathbf{q}$. Express, in terms of $\mathbf{p}$ and $\mathbf{q}$,

**(i)** $\overrightarrow{AC}$ **(ii)** $\overrightarrow{AD}$ **(iii)** $\overrightarrow{CD}$
**(iv)** $\overrightarrow{DE}$ **(v)** $\overrightarrow{EF}$ **(vi)** $\overrightarrow{BE}$

**SOLUTION**

**(i)** $\overrightarrow{AC} = \overrightarrow{AB} + \overrightarrow{BC}$
$= \mathbf{p} + \mathbf{q}$

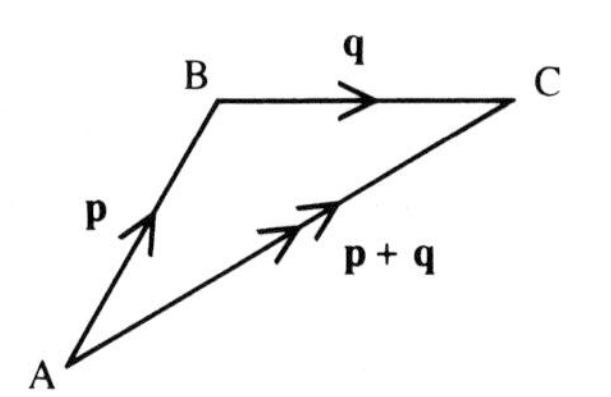


**(ii)** $\overrightarrow{AD} = 2\overrightarrow{BC}$
$= 2\mathbf{q}$

**(iii)** Since $\overrightarrow{AC} + \overrightarrow{CD} = \overrightarrow{AD}$

$\mathbf{p} + \mathbf{q} + \overrightarrow{CD} = 2\mathbf{q}$

and so $\overrightarrow{CD} = \mathbf{q} - \mathbf{p}$

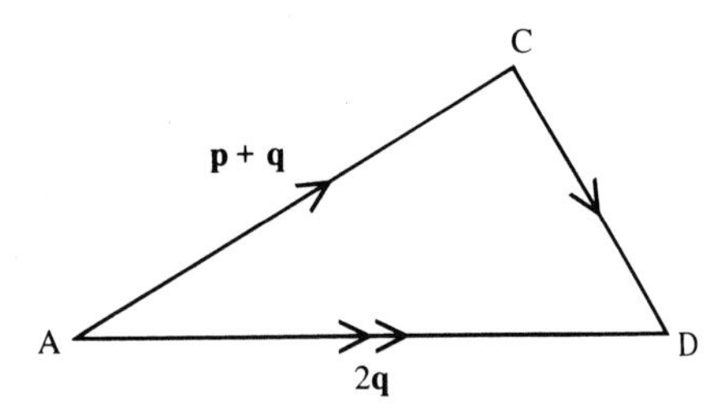


**(iv)** $\overrightarrow{DE} = -\overrightarrow{AB}$
$= -\mathbf{p}$

**(v)** $\overrightarrow{EF} = -\overrightarrow{BC}$
$= -\mathbf{q}$

**(vi)** $\overrightarrow{BE} = \overrightarrow{BC} + \overrightarrow{CD} + \overrightarrow{DE}$
$= \mathbf{q} + (\mathbf{q} - \mathbf{p}) + -\mathbf{p}$
$= 2\mathbf{q} - 2\mathbf{p}$

Notice that $\overrightarrow{BE} = 2\overrightarrow{CD}$.

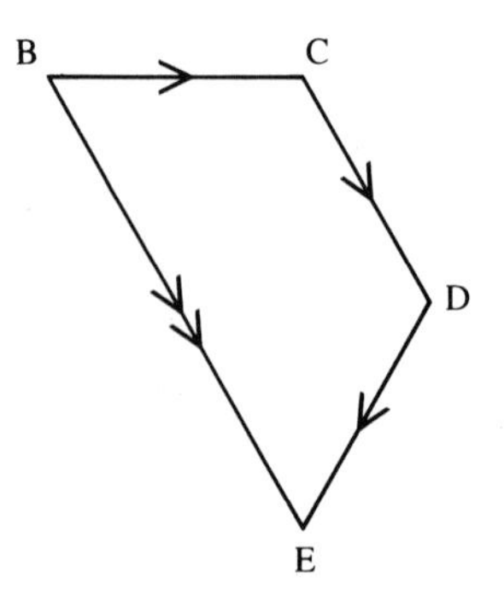


**Figure 5.17**

### Unit vectors

A unit vector is a vector with a magnitude of 1, like **i** and **j**. To find the unit vector in the same direction as a given vector, divide that vector by its magnitude.

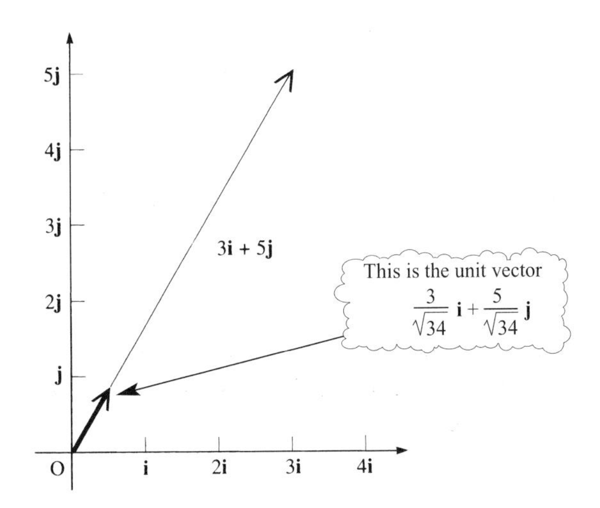


**Figure 5.18**

Thus the vector $3\mathbf{i} + 5\mathbf{j}$ (figure 5.18) has magnitude $\sqrt{3^2 + 5^2} = \sqrt{34}$, and so the vector $\frac{3}{\sqrt{34}}\mathbf{i} + \frac{5}{\sqrt{34}}\mathbf{j}$ is a unit vector. It has magnitude 1.

The unit vector in the direction of vector **a** is written as $\hat{\mathbf{a}}$ and read as 'a hat'.

**EXERCISE 5B**

**1** Simplify

**(i)** $\begin{pmatrix}2\\3\end{pmatrix} + \begin{pmatrix}4\\5\end{pmatrix}$ **(ii)** $\begin{pmatrix}2\\-1\end{pmatrix} + \begin{pmatrix}-1\\2\end{pmatrix}$

**(iii)** $\begin{pmatrix}3\\4\end{pmatrix} + \begin{pmatrix}-3\\-4\end{pmatrix}$ **(iv)** $3\begin{pmatrix}2\\1\end{pmatrix} + 2\begin{pmatrix}1\\-2\end{pmatrix}$

**(v)** $6(3\mathbf{i} - 2\mathbf{j}) - 9(2\mathbf{i} - \mathbf{j})$

**2** The vectors **p**, **q** and **r** are given by

$$\mathbf{p} = 3\mathbf{i} + 2\mathbf{j} \qquad \mathbf{q} = 2\mathbf{i} + 2\mathbf{j} \qquad \mathbf{r} = -3\mathbf{i} - \mathbf{j}.$$

Find, in component form, the vectors

**(i)** $\mathbf{p} + \mathbf{q} + \mathbf{r}$ **(ii)** $\mathbf{p} - \mathbf{q}$

**(iii)** $\mathbf{p} + \mathbf{r}$ **(iv)** $3(\mathbf{p} - \mathbf{q}) + 2(\mathbf{p} + \mathbf{r})$

**(v)** $4\mathbf{p} - 3\mathbf{q} + 2\mathbf{r}$

**3** In the diagram, PQRS is a parallelogram and $\overrightarrow{PQ} = \mathbf{a}$, $\overrightarrow{PS} = \mathbf{b}$.

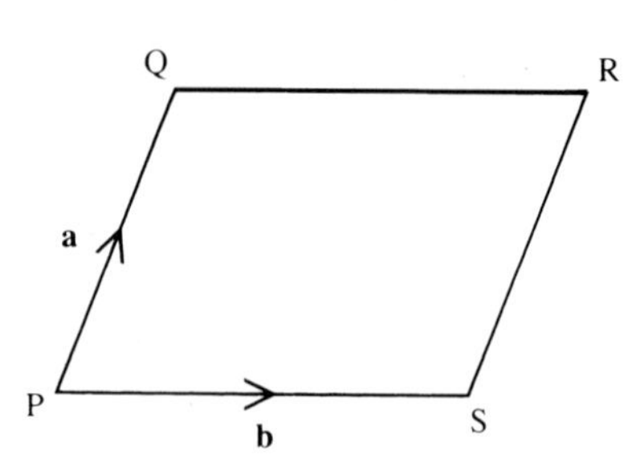


**(i)** Write, in terms of **a** and **b**, the following vectors.

**(a)** $\overrightarrow{QR}$ **(b)** $\overrightarrow{PR}$ **(c)** $\overrightarrow{QS}$

**(ii)** The mid-point of PR is M. Find **(a)** $\overrightarrow{PM}$, **(b)** $\overrightarrow{QM}$.

**(iii)** Explain why this shows you that the diagonals of a parallelogram bisect each other.

**4** In the diagram, ABCD is a kite. AC and BD meet at M.

$\overrightarrow{AB} = \mathbf{i} + \mathbf{j}$ and $\overrightarrow{AD} = \mathbf{i} - 2\mathbf{j}$

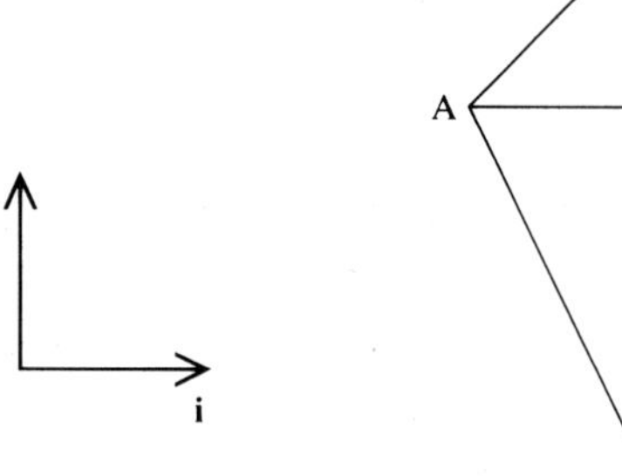


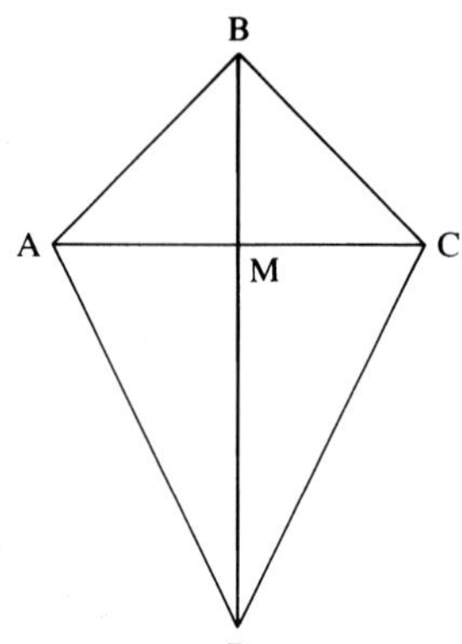


**(i)** Use the facts that the diagonals of a kite meet at right angles and that M is the mid-point of AC to find, in terms of **i** and **j**

**(a)** $\overrightarrow{AM}$ **(b)** $\overrightarrow{AC}$ **(c)** $\overrightarrow{BC}$ **(d)** $\overrightarrow{CD}$.

**(ii)** Verify that $|\overrightarrow{AB}| = |\overrightarrow{BC}|$ and $|\overrightarrow{AD}| = |\overrightarrow{CD}|$.

**5** In the diagram, ABC is a triangle. L, M and N are the mid-points of the sides BC, CA and AB.

$\overrightarrow{AB} = \mathbf{p}$ and $\overrightarrow{AC} = \mathbf{q}$

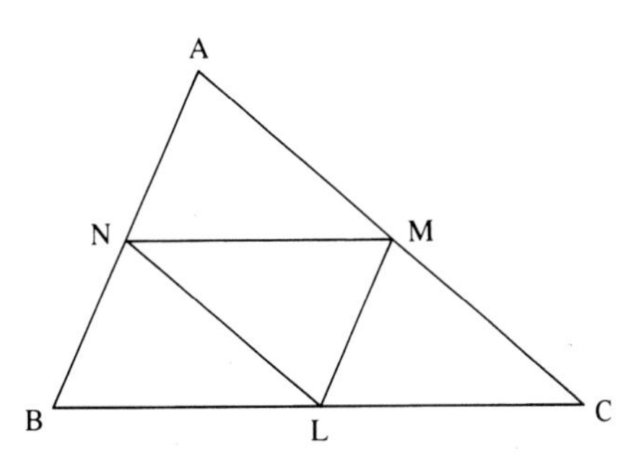


**(i)** Find, in terms of **p** and **q**, $\overrightarrow{BC}$, $\overrightarrow{MN}$, $\overrightarrow{LM}$ and $\overrightarrow{LN}$.

**(ii)** Explain how your results from part (i) show you that the sides of triangle LMN are parallel to those of triangle ABC, and half their lengths.

**6** Find unit vectors in the same directions as the following vectors.

**(i)** $\begin{pmatrix} 2 \\ 3 \end{pmatrix}$ **(ii)** $3\mathbf{i} + 4\mathbf{j}$ **(iii)** $\begin{pmatrix} -2 \\ -2 \end{pmatrix}$ **(iv)** $5\mathbf{i} - 12\mathbf{j}$

**(v)** $6\mathbf{i}$ **(vi)** $\begin{pmatrix} -2 \\ 4 \end{pmatrix}$ **(vii)** $\begin{pmatrix} -1 \\ 2 \end{pmatrix}$ **(viii)** $\begin{pmatrix} 3 \\ 6 \end{pmatrix}$

**(ix)** $\begin{pmatrix} r\cos\alpha \\ r\sin\alpha \end{pmatrix}$ **(x)** $\begin{pmatrix} 1 \\ \tan\beta \end{pmatrix}$

## Co-ordinate geometry using vectors: two dimensions

Two-dimensional co-ordinate geometry involves the study of points, given as co-ordinates, and lines, given as cartesian equations. The same work may also be treated using vectors.

The co-ordinates of a point, say (3, 4), are replaced by its position vector $\begin{pmatrix} 3 \\ 4 \end{pmatrix}$ or $3\mathbf{i} + 4\mathbf{j}$. The cartesian equation of a line is replaced by its vector form, and this is introduced on page 139.

Since most two-dimensional problems are readily solved using the methods of cartesian co-ordinate geometry, as introduced in *Pure Mathematics 1*, Chapter 1, why go to the trouble of relearning it all in vectors? The answer is that vector methods are very much easier to use in many three-dimensional situations than cartesian methods are. In preparation for that, we review some familiar two-dimensional work in this section, comparing cartesian and vector methods.

## The vector joining two points

In figure 5.19, start by looking at two points A(2, –1) and B(4, 3); that is the points with position vectors $\overrightarrow{OA} = \begin{pmatrix} 2 \\ -1 \end{pmatrix}$ and $\overrightarrow{OB} = \begin{pmatrix} 4 \\ 3 \end{pmatrix}$, alternatively $2\mathbf{i} - \mathbf{j}$ and $4\mathbf{i} + 3\mathbf{j}$.

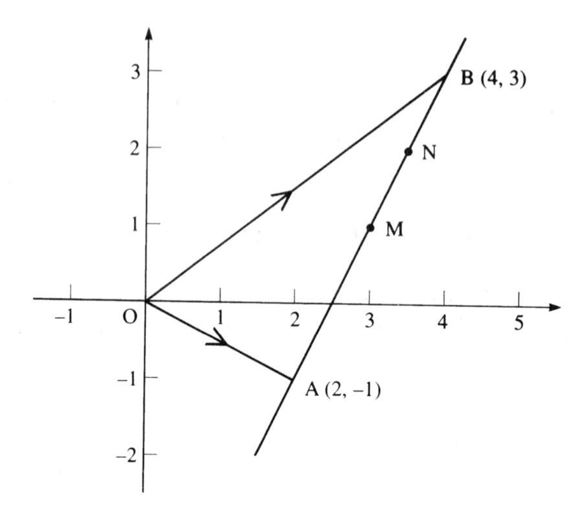


**Figure 5.19**

The vector joining A to B is $\overrightarrow{AB}$ and this is given by

$$\begin{aligned}\overrightarrow{AB} &= \overrightarrow{AO} + \overrightarrow{OB} \\ &= -\overrightarrow{OA} + \overrightarrow{OB} \\ &= \overrightarrow{OB} - \overrightarrow{OA} \\ &= \begin{pmatrix} 4 \\ 3 \end{pmatrix} - \begin{pmatrix} 2 \\ -1 \end{pmatrix} = \begin{pmatrix} 2 \\ 4 \end{pmatrix}.\end{aligned}$$

Since $\overrightarrow{AB} = \begin{pmatrix} 2 \\ 4 \end{pmatrix}$, then it follows that the length of AB is given by

$$\begin{aligned}|\overrightarrow{AB}| &= \sqrt{2^2 + 4^2} \\ &= \sqrt{20}\end{aligned}$$

You can find the position vectors of points along AB as follows.

The mid-point, M, has position vector $\overrightarrow{OM}$, given by

$$\begin{aligned}\overrightarrow{OM} &= \overrightarrow{OA} + \tfrac{1}{2}\overrightarrow{AB} \\ &= \begin{pmatrix} 2 \\ -1 \end{pmatrix} + \tfrac{1}{2}\begin{pmatrix} 2 \\ 4 \end{pmatrix} \\ &= \begin{pmatrix} 3 \\ 1 \end{pmatrix}.\end{aligned}$$

In the same way, the position vector of the point N, three-quarters of the distance from A to B, is given by

$$\overrightarrow{ON} = \begin{pmatrix} 2 \\ -1 \end{pmatrix} + \tfrac{3}{4}\begin{pmatrix} 2 \\ 4 \end{pmatrix}$$

$$= \begin{pmatrix} 3\tfrac{1}{2} \\ 2 \end{pmatrix}$$

and it is possible to find the position vector of any other point of subdivision of the line AB in the same way.

## The vector equation of a line

It is now a small step to go from finding the position vector of any point on the line AB to finding the vector form of the equation of the line AB. To take this step, you will find it helpful to carry out the following activity.

**ACTIVITY**

The position vectors of a set of points are given by

$$\mathbf{r} = \begin{pmatrix} 2 \\ -1 \end{pmatrix} + \lambda \begin{pmatrix} 2 \\ 4 \end{pmatrix}$$

where $\lambda$ is a parameter which may take any value.

**(i)** Show that $\lambda = 2$ corresponds to the point with position vector $\begin{pmatrix} 6 \\ 7 \end{pmatrix}$.

**(ii)** Find the position vectors of points corresponding to values of $\lambda$ of $-2, -1, 0, \tfrac{1}{2}, \tfrac{3}{4}, 1, 3$.

**(iii)** Mark all your points on a sheet of graph paper and show that when they are joined up they give the line AB in figure 5.19.

**(iv)** State what values of $\lambda$ correspond to the points A, B, M and N.

**(v)** What can you say about the position of the point if

**(a)** $0 < \lambda < 1$?

**(b)** $\lambda > 1$?

**(c)** $\lambda < 0$?

### CONCLUSIONS FROM THE ACTIVITY

This activity should have convinced you that

$$\mathbf{r} = \begin{pmatrix} 2 \\ -1 \end{pmatrix} + \lambda \begin{pmatrix} 2 \\ 4 \end{pmatrix}$$

is the equation of the line passing through $(2, -1)$ and $(4, 3)$, written in vector form.

You may find it helpful to think of this in these terms:

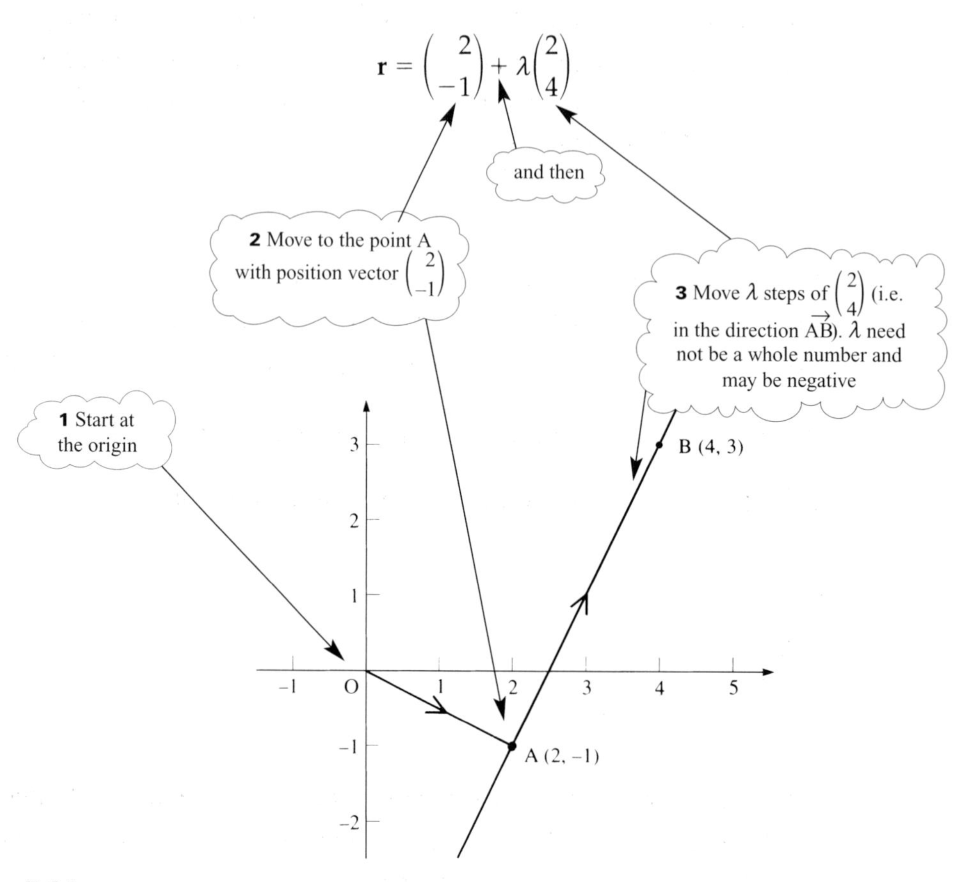


**Figure 5.20**

You should also have noticed that:

| | |
|---|---|
| $\lambda = 0$ | corresponds to the point A |
| $\lambda = 1$ | corresponds to the point B |
| $0 < \lambda < 1$ | the point lies between A and B |
| $\lambda > 1$ | the point lies beyond B |
| $\lambda < 0$ | the point lies beyond A. |

The vector form of the equation is not unique; there are many (in fact infinitely many) different ways in which the equation of any particular line may be expressed. There are two reasons for this: direction and location.

## DIRECTION

The direction of the line in the example is $\begin{pmatrix}2\\4\end{pmatrix}$. That means that for every 2 units along (in the **i** direction), the line goes up 4 units (in the **j** direction). This is equivalent to stating that for every 1 unit along, the line goes up 2 units, corresponding to the equation

$$\mathbf{r} = \begin{pmatrix}2\\-1\end{pmatrix} + \lambda\begin{pmatrix}1\\2\end{pmatrix}.$$

The only difference is that the two equations have different values of $\lambda$ for particular points. In the first equation, point B, with position vector $\begin{pmatrix}4\\3\end{pmatrix}$, corresponds to a value of $\lambda$ of 1. In the second equation, the value of $\lambda$ for B is 2.

The direction $\begin{pmatrix}2\\4\end{pmatrix}$ is the same as $\begin{pmatrix}1\\2\end{pmatrix}$, or as any multiple of $\begin{pmatrix}1\\2\end{pmatrix}$ such as $\begin{pmatrix}3\\6\end{pmatrix}$, $\begin{pmatrix}-5\\-10\end{pmatrix}$ or $\begin{pmatrix}100.5\\201\end{pmatrix}$. Any of these could be used in the vector equation of the line.

## LOCATION

In the equation

$$\mathbf{r} = \begin{pmatrix}2\\-1\end{pmatrix} + \lambda\begin{pmatrix}2\\4\end{pmatrix}$$

$\begin{pmatrix}2\\-1\end{pmatrix}$ is the position vector of the point A on the line, and represents the point at which the line was joined. However, this could have been any other point on the line, such as M(3, 1), B(4, 3) etc. Consequently

$$\mathbf{r} = \begin{pmatrix}3\\1\end{pmatrix} + \lambda\begin{pmatrix}2\\4\end{pmatrix}$$

and

$$\mathbf{r} = \begin{pmatrix}4\\3\end{pmatrix} + \lambda\begin{pmatrix}2\\4\end{pmatrix}$$

are also equations of the same line, and there are infinitely many other possibilities, one corresponding to each point on the line.

***Notes***

**(i)** It is usual to refer to any valid vector form of the equation as *the* vector equation of the line even though it is not unique.

**(ii)** It is often a good idea to give the direction vector in its simplest integer form: for example, replacing $\begin{pmatrix}2\\4\end{pmatrix}$ with $\begin{pmatrix}1\\2\end{pmatrix}$.

### The general vector form of the equation of a line

If A and B are points with position **a** and **b**, then the equation

$$\mathbf{r} = \overrightarrow{OA} + \lambda\overrightarrow{AB}$$

may be written as $\mathbf{r} = \mathbf{a} + \lambda(\mathbf{b} - \mathbf{a})$

which implies $\mathbf{r} = (1 - \lambda)\mathbf{a} + \lambda\mathbf{b}$.

This is the general vector form of the equation of the line joining two points.

**ACTIVITY**

Plot the following lines on the same sheet of graph paper. When you have done so, explain why certain among them are the same as each other, others are parallel to each other, and others are in different directions.

**(i)** $\mathbf{r} = \begin{pmatrix} 2 \\ -1 \end{pmatrix} + \lambda\begin{pmatrix} 1 \\ 2 \end{pmatrix}$ **(ii)** $\mathbf{r} = \begin{pmatrix} 2 \\ -1 \end{pmatrix} + \lambda\begin{pmatrix} -1 \\ 2 \end{pmatrix}$ **(iii)** $\mathbf{r} = \begin{pmatrix} 0 \\ 2 \end{pmatrix} + \lambda\begin{pmatrix} 1 \\ 2 \end{pmatrix}$

**(iv)** $\mathbf{r} = \begin{pmatrix} 1 \\ -3 \end{pmatrix} + \lambda\begin{pmatrix} 3 \\ 6 \end{pmatrix}$ **(v)** $\mathbf{r} = \begin{pmatrix} 4 \\ 3 \end{pmatrix} + \lambda\begin{pmatrix} 1 \\ -2 \end{pmatrix}$

### Cartesian and vector forms of the equation of a line

To find the cartesian form of the equation of a line which is given in vector form

$$\mathbf{r} = \begin{pmatrix} 2 \\ -1 \end{pmatrix} + \lambda\begin{pmatrix} 2 \\ 4 \end{pmatrix}$$

write **r** as $\begin{pmatrix} x \\ y \end{pmatrix}$, so the equation of the line becomes

$$\begin{pmatrix} x \\ y \end{pmatrix} = \begin{pmatrix} 2 \\ -1 \end{pmatrix} + \lambda\begin{pmatrix} 2 \\ 4 \end{pmatrix}$$

or $x = 2 + 2\lambda$

$y = -1 + 4\lambda$

The last two equations can be rewritten as

$$\frac{x-2}{2} = \lambda \quad \text{and} \quad \frac{y+1}{4} = \lambda$$

$$\Rightarrow \quad \frac{x-2}{2} = \frac{y+1}{4} \; (= \lambda).$$

The equation is now in cartesian form and may be tidied up to give $y = 2x - 5$.

When converting from cartesian form to vector form, you need (a) to find any point on the line, and (b) to convert the gradient into a vector with the same direction, as shown in the following example.

**EXAMPLE 5.9**

Write $y = \frac{1}{3}x + 2$ in vector form.

**SOLUTION**

First find any point on the line. For example, when $x = 0$, $y = 2$ and so the point (0, 2) with position vector $\begin{pmatrix} 0 \\ 2 \end{pmatrix}$ is on the line. Then convert the gradient into a vector with the same direction. The equation of the line is of the form $y = mx + c$ and so its gradient $m$ is $\frac{1}{3}$.

The vector $\begin{pmatrix} 3 \\ 1 \end{pmatrix}$ has gradient $\frac{1}{3}$.

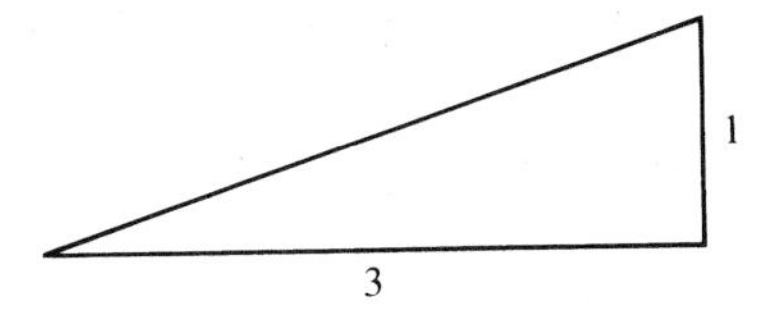


**Figure 5.21**

So the vector equation of the line is

$$\mathbf{r} = \begin{pmatrix} 0 \\ 2 \end{pmatrix} + \lambda\begin{pmatrix} 3 \\ 1 \end{pmatrix}.$$

---

 Remember that there are other ways of writing this vector equation.

---

### The intersection of two lines

**EXAMPLE 5.10**

Find the position vector of the point where the following lines intersect:

$$\mathbf{r} = \begin{pmatrix} 2 \\ 3 \end{pmatrix} + \lambda\begin{pmatrix} 1 \\ 2 \end{pmatrix} \quad \text{and} \quad \mathbf{r} = \begin{pmatrix} 6 \\ 1 \end{pmatrix} + \mu\begin{pmatrix} 1 \\ -3 \end{pmatrix}.$$

Note here that different letters are used for the parameters in the two equations to avoid confusion.

**SOLUTION**

When the lines intersect, the position vector is the same for each of them:

$$\mathbf{r} = \begin{pmatrix} x \\ y \end{pmatrix} = \begin{pmatrix} 2 \\ 3 \end{pmatrix} + \lambda\begin{pmatrix} 1 \\ 2 \end{pmatrix} = \begin{pmatrix} 6 \\ 1 \end{pmatrix} + \mu\begin{pmatrix} 1 \\ -3 \end{pmatrix}.$$

This gives two simultaneous equations for $\lambda$ and $\mu$:

$$x: \quad 2 + \lambda = 6 + \mu \quad \Rightarrow \quad \lambda - \mu = 4$$

$$y: \quad 3 + 2\lambda = 1 - 3\mu \quad \Rightarrow \quad 2\lambda + 3\mu = -2.$$

Solving these gives $\lambda = 2$ and $\mu = -2$. Substituting in either equation gives

$$\mathbf{r} = \begin{pmatrix} 4 \\ 7 \end{pmatrix}$$

which is the position vector of the point of intersection.

**EXAMPLE 5.11**

Find the co-ordinates of the point of intersection of the lines joining A(1, 6) to B(4, 0), and C(1, 1) to D(5, 3).

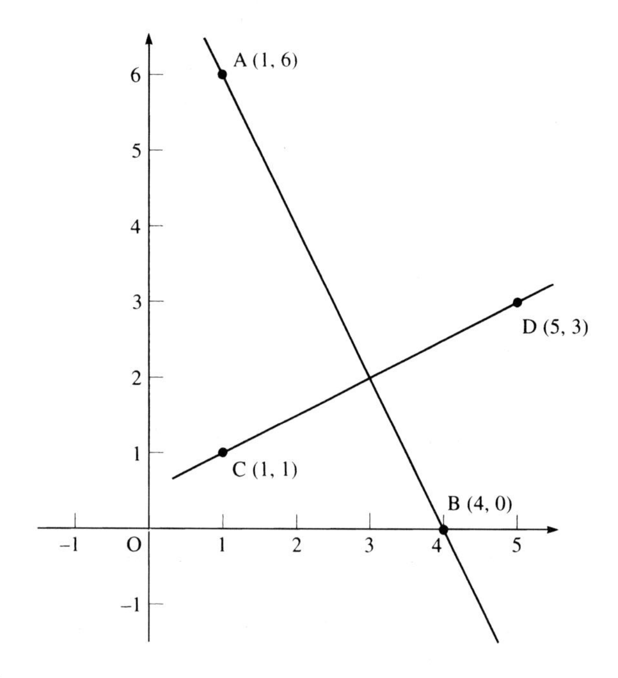


**Figure 5.22**

**SOLUTION**

$$\overrightarrow{AB} = \begin{pmatrix} 4 \\ 0 \end{pmatrix} - \begin{pmatrix} 1 \\ 6 \end{pmatrix} = \begin{pmatrix} 3 \\ -6 \end{pmatrix}$$

and so the vector equation of line AB is

$$\mathbf{r} = \overrightarrow{OA} + \lambda\overrightarrow{AB}$$

$$\mathbf{r} = \begin{pmatrix} 1 \\ 6 \end{pmatrix} + \lambda\begin{pmatrix} 3 \\ -6 \end{pmatrix}$$

$$\overrightarrow{CD} = \begin{pmatrix} 5 \\ 3 \end{pmatrix} - \begin{pmatrix} 1 \\ 1 \end{pmatrix} = \begin{pmatrix} 4 \\ 2 \end{pmatrix}$$

and so the vector equation of line CD is

$$\mathbf{r} = \overrightarrow{OC} + \mu\overrightarrow{CD}$$

$$\mathbf{r} = \begin{pmatrix}1\\1\end{pmatrix} + \mu\begin{pmatrix}4\\2\end{pmatrix}.$$

The intersection of these lines is at

$$\mathbf{r} = \begin{pmatrix}1\\6\end{pmatrix} + \lambda\begin{pmatrix}3\\-6\end{pmatrix} = \begin{pmatrix}1\\1\end{pmatrix} + \mu\begin{pmatrix}4\\2\end{pmatrix}.$$

$x$: $\quad 1 + 3\lambda = 1 + 4\mu \;\Rightarrow\; 3\lambda - 4\mu = 0$ ①

$y$: $\quad 6 - 6\lambda = 1 + 2\mu \;\Rightarrow\; 6\lambda + 2\mu = 5$ ②

Solve ① and ② simultaneously:

①: $\quad 3\lambda - 4\mu = 0$
② × 2: $\quad 12\lambda + 4\mu = 10$
Add: $\quad 15\lambda = 10$

$\Rightarrow \quad \lambda = \frac{2}{3}.$

Substitute $\lambda = \frac{2}{3}$ in the equation for AB:

$$\Rightarrow \quad \mathbf{r} = \begin{pmatrix}1\\6\end{pmatrix} + \tfrac{2}{3}\begin{pmatrix}3\\-6\end{pmatrix}$$

$$\Rightarrow \quad \mathbf{r} = \begin{pmatrix}3\\2\end{pmatrix}.$$

The point of intersection has co-ordinates (3, 2).

***Note***

Alternatively, you could have found $\mu = \frac{1}{2}$ and substituted in the equation for CD.

**EXERCISE 5C**

**1** For each of these pairs of points, A and B, write down:
**(i)** the vector $\overrightarrow{AB}$
**(ii)** $|\overrightarrow{AB}|$
**(iii)** the position vector of the mid-point of AB.
**(a)** A is (2, 3), B is (4, 11).
**(b)** A is (4, 3), B is (0, 0).
**(c)** A is (–2, –1), B is (4, 7).
**(d)** A is (–3, 4), B is (3, –4).
**(e)** A is (–10, –8), B is (–5, 4).

**2** Find the equation of each of these lines in vector form.

**(i)** Joining (2, 1) to (4, 5).
**(ii)** Joining (3, 5) to (0, 8).
**(iii)** Joining (–6, –6) to (4, 4).
**(iv)** Through (5, 3) in the same direction as $\mathbf{i} + \mathbf{j}$.
**(v)** Through (2, 1) parallel to $6\mathbf{i} + 3\mathbf{j}$.
**(vi)** Through (0, 0) parallel to $\begin{pmatrix} -1 \\ 4 \end{pmatrix}$.
**(vii)** Joining (0, 0) to (–2, 8).
**(viii)** Joining (3, –12) to (–1, 4).

**3** Write these lines in cartesian form.

**(i)** $\mathbf{r} = \begin{pmatrix} 1 \\ 2 \end{pmatrix} + \lambda\begin{pmatrix} 1 \\ 3 \end{pmatrix}$
**(ii)** $\mathbf{r} = \begin{pmatrix} -2 \\ 0 \end{pmatrix} + \lambda\begin{pmatrix} -2 \\ -1 \end{pmatrix}$
**(iii)** $\mathbf{r} = \begin{pmatrix} 1 \\ 0 \end{pmatrix} + \lambda\begin{pmatrix} 4 \\ 4 \end{pmatrix}$
**(iv)** $\mathbf{r} = \begin{pmatrix} 4 \\ 3 \end{pmatrix} + \lambda\begin{pmatrix} 1 \\ 1 \end{pmatrix}$
**(v)** $\mathbf{r} = \begin{pmatrix} 2 \\ 5 \end{pmatrix} + \lambda\begin{pmatrix} 4 \\ 0 \end{pmatrix}$

**4** Write these lines in vector form.

**(i)** $y = 2x + 3$
**(ii)** $y = x - 4$
**(iii)** $y = \frac{1}{2}x - 1$
**(iv)** $y = -\frac{1}{4}x$
**(v)** $x + 2y = 8$

**5** Find the position vector of the point of intersection of each of these pairs of lines.

**(i)** $\mathbf{r} = \begin{pmatrix} 2 \\ 1 \end{pmatrix} + \lambda\begin{pmatrix} 1 \\ 0 \end{pmatrix}$ : $\mathbf{r} = \begin{pmatrix} 3 \\ 0 \end{pmatrix} + \mu\begin{pmatrix} 1 \\ 1 \end{pmatrix}$
**(ii)** $\mathbf{r} = \begin{pmatrix} 2 \\ -1 \end{pmatrix} + \lambda\begin{pmatrix} 1 \\ 2 \end{pmatrix}$ : $\mathbf{r} = \mu\begin{pmatrix} 1 \\ 1 \end{pmatrix}$
**(iii)** $\mathbf{r} = \begin{pmatrix} 0 \\ 5 \end{pmatrix} + \lambda\begin{pmatrix} -2 \\ -2 \end{pmatrix}$ : $\mathbf{r} = \begin{pmatrix} 0 \\ -7 \end{pmatrix} + \mu\begin{pmatrix} 1 \\ 2 \end{pmatrix}$
**(iv)** $\mathbf{r} = \begin{pmatrix} -2 \\ -3 \end{pmatrix} + \lambda\begin{pmatrix} -1 \\ 3 \end{pmatrix}$ : $\mathbf{r} = \begin{pmatrix} 1 \\ 3 \end{pmatrix} + \mu\begin{pmatrix} 2 \\ -1 \end{pmatrix}$
**(v)** $\mathbf{r} = \begin{pmatrix} 2 \\ 7 \end{pmatrix} + \lambda\begin{pmatrix} 1 \\ -1 \end{pmatrix}$ : $\mathbf{r} = \begin{pmatrix} 5 \\ 1 \end{pmatrix} + \mu\begin{pmatrix} 1 \\ 2 \end{pmatrix}$

**6** In this question the origin is taken to be at a harbour and the unit vectors **i** and **j** to have lengths of 1 km in the directions E and N.

A cargo vessel leaves the harbour and its position vector $t$ hours later is given by

$$\mathbf{r}_1 = 12t\mathbf{i} + 16t\mathbf{j}.$$

A fishing boat is trawling nearby and its position at time $t$ is given by

$$\mathbf{r}_2 = (10 - 3t)\mathbf{i} + (8 + 4t)\mathbf{j}.$$

**(i)** How far apart are the two boats when the cargo vessel leaves harbour?
**(ii)** How fast is each boat travelling?
**(iii)** What happens?

**7** The points A(1, 0), B(7, 2) and C(13, 7) are the vertices of a triangle. The mid-points of the sides BC, CA and AB are L, M and N.

**(i)** Write down the position vectors of L, M and N.

**(ii)** Find the vector equations of the lines AL, BM and CN.

**(iii)** Find the intersections of these pairs of lines:

**(a)** AL and BM **(b)** BM and CN

**(iv)** What do you notice?

## The angle between two vectors

To find the angle $\theta$ between the two vectors

$$\overrightarrow{OA} = \mathbf{a} = a_1\mathbf{i} + a_2\mathbf{j} \quad \text{and} \quad \overrightarrow{OB} = \mathbf{b} = b_1\mathbf{i} + b_2\mathbf{j}$$

start by applying the cosine rule to triangle OAB (figure 5.23):

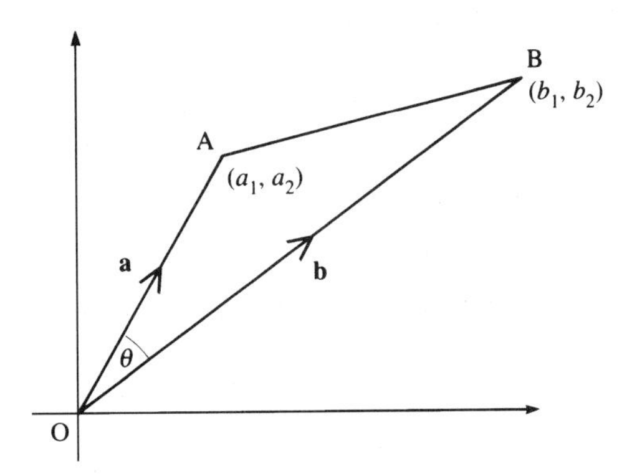


**Figure 5.23**

$$\cos\theta = \frac{OA^2 + OB^2 - AB^2}{2OA \times OB}.$$

In this, OA, OB and AB are the lengths of the vectors $\overrightarrow{OA}$, $\overrightarrow{OB}$ and $\overrightarrow{AB}$, and so

$$OA = |\mathbf{a}| = \sqrt{a_1^2 + a_2^2} \quad \text{and} \quad OB = |\mathbf{b}| = \sqrt{b_1^2 + b_2^2}$$

The vector $\overrightarrow{AB} = \mathbf{b} - \mathbf{a}$

$$= (b_1\mathbf{i} + b_2\mathbf{j}) - (a_1\mathbf{i} + a_2\mathbf{j})$$

$$= (b_1 - a_1)\mathbf{i} + (b_2 - a_2)\mathbf{j}$$

and so its length is given by

$$AB = |\mathbf{b} - \mathbf{a}| = \sqrt{(b_1 - a_1)^2 + (b_2 - a_2)^2}.$$

Substituting for OA, OB and AB in the cosine rule gives

$$\cos\theta = \frac{(a_1^2 + a_2^2) + (b_1^2 + b_2^2) - [(b_1 - a_1)^2 + (b_2 - a_2)^2]}{2\sqrt{a_1^2 + a_2^2}\,\sqrt{b_1^2 + b_2^2}}$$

$$= \frac{a_1^2 + a_2^2 + b_1^2 + b_2^2 - (b_1^2 - 2a_1b_1 + a_1^2 + b_2^2 - 2a_2b_2 + a_2^2)}{2\,|\mathbf{a}|\,|\mathbf{b}|}$$

This simplifies to

$$\cos\theta = \frac{2a_1b_1 + 2a_2b_2}{2\,|\mathbf{a}|\,|\mathbf{b}|}$$

$$= \frac{a_1b_1 + a_2b_2}{|\mathbf{a}|\,|\mathbf{b}|}.$$

The expression on the top line, $a_1b_1 + a_2b_2$, is called the *scalar product* (or *dot product*) of the vectors **a** and **b** and is written **a.b**. Thus

$$\cos\theta = \frac{\mathbf{a.b}}{|\mathbf{a}|\,|\mathbf{b}|}.$$

This result is usually written in the form

$$\mathbf{a.b} = |\mathbf{a}|\,|\mathbf{b}|\cos\theta.$$

The next example shows you how to use it to find the angle between two vectors given numerically.

**EXAMPLE 5.12**

Find the angle between the vectors $\begin{pmatrix} 3 \\ 4 \end{pmatrix}$ and $\begin{pmatrix} 5 \\ -12 \end{pmatrix}$.

**SOLUTION**

Let $\mathbf{a} = \begin{pmatrix} 3 \\ 4 \end{pmatrix} \Rightarrow |\mathbf{a}| = \sqrt{3^2 + 4^2} = 5$

and $\mathbf{b} = \begin{pmatrix} 5 \\ -12 \end{pmatrix} \Rightarrow |\mathbf{b}| = \sqrt{5^2 + (-12)^2} = 13.$

The scalar product

$$\begin{pmatrix} 3 \\ 4 \end{pmatrix}.\begin{pmatrix} 5 \\ -12 \end{pmatrix} = 3 \times 5 + 4 \times (-12)$$

$$= 15 - 48$$

$$= -33.$$

Substituting in $\mathbf{a.b} = |\mathbf{a}|\,|\mathbf{b}|\cos\theta$ gives

$$-33 = 5 \times 13 \times \cos\theta$$

$$\cos\theta = \frac{-33}{65}$$

$$\theta = 120.5°.$$

## Perpendicular vectors

Since $\cos 90° = 0$, it follows that if vectors **a** and **b** are perpendicular then **a.b** = 0.

Conversely, if the scalar product of two non-zero vectors is 0, they are perpendicular.

**EXAMPLE 5.13**

Show that the vectors $\mathbf{a} = \begin{pmatrix} 2 \\ 4 \end{pmatrix}$ and $\mathbf{b} = \begin{pmatrix} 6 \\ -3 \end{pmatrix}$ are perpendicular.

**SOLUTION**

The scalar product of the vectors is

$$\begin{aligned} \mathbf{a.b} &= \begin{pmatrix} 2 \\ 4 \end{pmatrix} \cdot \begin{pmatrix} 6 \\ -3 \end{pmatrix} \\ &= 2 \times 6 + 4 \times (-3) \\ &= 12 - 12 = 0. \end{aligned}$$

Therefore the vectors are perpendicular.

## Further points concerning the scalar product

- You will notice that the scalar product of two vectors is an ordinary number. It has size but no direction and so is a scalar, rather than a vector. It is for this reason that it is called the scalar product. There is another way of multiplying vectors that gives a vector as the answer; it is called the *vector product*. You will meet this in *Pure Mathematics 4*.

- The scalar product is calculated in the same way for three-dimensional vectors. For example:

$$\begin{pmatrix} 2 \\ 3 \\ 4 \end{pmatrix} \cdot \begin{pmatrix} 5 \\ 6 \\ 7 \end{pmatrix} = 2 \times 5 + 3 \times 6 + 4 \times 7 = 56.$$

  In general

$$\begin{pmatrix} a_1 \\ a_2 \\ a_3 \end{pmatrix} \cdot \begin{pmatrix} b_1 \\ b_2 \\ b_3 \end{pmatrix} = a_1b_1 + a_2b_2 + a_3b_3.$$

- The scalar product of two vectors is commutative. It has the same value whichever of them is on the left-hand side or right-hand side. Thus **a.b** = **b.a**, as in the following example.

$$\begin{pmatrix} 2 \\ 3 \end{pmatrix} \cdot \begin{pmatrix} 6 \\ 7 \end{pmatrix} = 2 \times 6 + 3 \times 7 = 33$$

$$\begin{pmatrix} 6 \\ 7 \end{pmatrix} \cdot \begin{pmatrix} 2 \\ 3 \end{pmatrix} = 6 \times 2 + 7 \times 3 = 33.$$

**EXERCISE 5D**

**1** Find the angles between these vectors.

**(i)** $2\mathbf{i} + 3\mathbf{j}$ and $4\mathbf{i} + \mathbf{j}$ **(ii)** $2\mathbf{i} - \mathbf{j}$ and $\mathbf{i} + 2\mathbf{j}$

**(iii)** $\begin{pmatrix} -1 \\ -1 \end{pmatrix}$ and $\begin{pmatrix} -1 \\ -2 \end{pmatrix}$ **(iv)** $4\mathbf{i} + \mathbf{j}$ and $\mathbf{i} + \mathbf{j}$

**(v)** $\begin{pmatrix} 2 \\ 3 \end{pmatrix}$ and $\begin{pmatrix} -6 \\ 4 \end{pmatrix}$ **(vi)** $\begin{pmatrix} 3 \\ -1 \end{pmatrix}$ and $\begin{pmatrix} -6 \\ 2 \end{pmatrix}$

**2** Points A, B, C and D are (1, 0), (9, 4), (6, 1) and (9, 7), respectively.

**(i)** Write down the vector equation of line AB.

**(ii)** Write down the vector equation of line CD.

**(iii)** Find the position vector of the point of intersection.

**(iv)** Find the angle between the lines AB and CD.

**3** The equations of the four sides AB, BC, CD, DA of a quadrilateral are:

AB: $\mathbf{r} = \begin{pmatrix} 1 \\ 1 \end{pmatrix} + \lambda_1 \begin{pmatrix} 4 \\ 1 \end{pmatrix}$ BC: $\mathbf{r} = \begin{pmatrix} 1 \\ 1 \end{pmatrix} + \lambda_2 \begin{pmatrix} 1 \\ 3 \end{pmatrix}$

CD: $\mathbf{r} = \begin{pmatrix} 6 \\ 5 \end{pmatrix} + \lambda_3 \begin{pmatrix} 4 \\ 1 \end{pmatrix}$ DA: $\mathbf{r} = \begin{pmatrix} 6 \\ 5 \end{pmatrix} + \lambda_4 \begin{pmatrix} 1 \\ 3 \end{pmatrix}$.

**(i)** Look carefully at the equations of the four lines and state, with reasons, what sort of quadrilateral it is.

**(ii)** Find the co-ordinates of the four vertices of the quadrilateral.

**(iii)** Find the internal angles of the quadrilateral.

**4** The points A, B and C have co-ordinates (3, 2), (6, 3) and (5, 6), respectively.

**(i)** Write down the vectors $\overrightarrow{AB}$ and $\overrightarrow{BC}$.

**(ii)** Show that the angle ABC is 90°.

**(iii)** Show that $|\overrightarrow{AB}| = |\overrightarrow{BC}|$.

The figure ABCD is a square.

**(iv)** Find the co-ordinates of the point D.

**5** Three points P, Q and R have position vectors, $\mathbf{p}$, $\mathbf{q}$ and $\mathbf{r}$ respectively, where

$$\mathbf{p} = 7\mathbf{i} + 10\mathbf{j}, \quad \mathbf{q} = 3\mathbf{i} + 12\mathbf{j}, \quad \mathbf{r} = -\mathbf{i} + 4\mathbf{j}.$$

**(i)** Write down the vectors $\overrightarrow{PQ}$ and $\overrightarrow{RQ}$, and show that they are perpendicular.

**(ii)** Using a scalar product, or otherwise, find the angle PRQ.

**(iii)** Find the position vector of S, the mid-point of PR.

**(iv)** Show that $|\overrightarrow{QS}| = |\overrightarrow{RS}|$. Using your previous results, or otherwise, find the angle PSQ.

[MEI]

# Co-ordinate geometry using vectors: three dimensions

## Points

In three dimensions, a point has 3 co-ordinates, usually called $x$, $y$ and $z$.

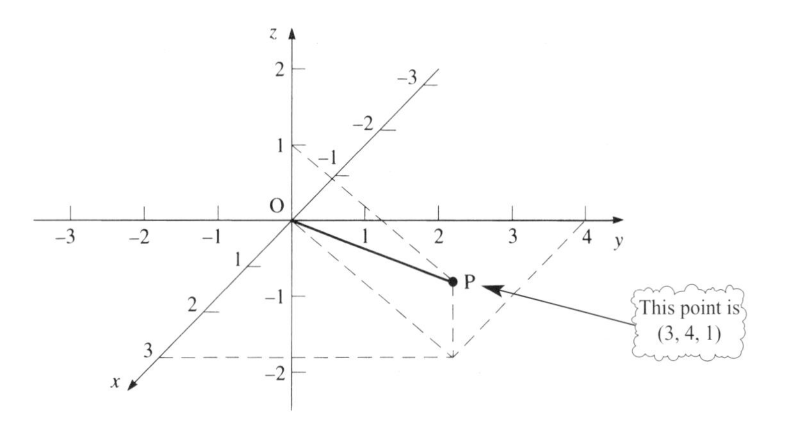


**Figure 5.24**

The axes are conventionally arranged as shown in figure 5.24, where the point P is (3, 4, 1). Even on correctly drawn three-dimensional grids, it is often hard to see the relationship between the points, lines and planes, so it is seldom worth your while trying to plot points accurately.

ACTIVITY

There are many important results that can be extended from two dimensions into three dimensions. In this activity you are asked to prove two of these.

### The length of a vector

In two dimensions, the use of Pythagoras' theorem leads to the result that a vector $a_1\mathbf{i} + a_2\mathbf{j}$ has length $|\mathbf{a}|$ given by

$$|\mathbf{a}| = \sqrt{a_1^2 + a_2^2}.$$

Show that the length of the three-dimensional vector $a_1\mathbf{i} + a_2\mathbf{j} + a_3\mathbf{k}$ is given by

$$|\mathbf{a}| = \sqrt{a_1^2 + a_2^2 + a_3^2}.$$

### The angle between two vectors

The angle $\theta$ between the vectors $\mathbf{a} = a_1\mathbf{i} + a_2\mathbf{j}$ and $b_1\mathbf{i} + b_2\mathbf{j}$ in two dimensions is given by

$$\cos\theta = \frac{a_1b_1 + a_2b_2}{\sqrt{a_1^2 + a_2^2}\sqrt{b_1^2 + b_2^2}} = \frac{\mathbf{a.b}}{|\mathbf{a}|\,|\mathbf{b}|}$$

where **a.b** is the scalar product of **a** and **b**. This result was proved by using the cosine rule on page 147–148.

Show that the angle between the three-dimensional vectors

$$\mathbf{a} = a_1\mathbf{i} + a_2\mathbf{j} + a_3\mathbf{k} \quad \text{and} \quad \mathbf{b} = b_1\mathbf{i} + b_2\mathbf{j} + b_3\mathbf{k}$$

is also given by

$$\cos\theta = \frac{\mathbf{a.b}}{|\mathbf{a}|\,|\mathbf{b}|}$$

but the scalar product **a.b** is now

$$\mathbf{a.b} = a_1b_1 + a_2b_2 + a_3b_3.$$

### Vectors

The position vector of the point P in figure 5.24 is given by

$$3\mathbf{i} + 4\mathbf{j} + \mathbf{k} \quad \text{or} \quad \begin{pmatrix} 3 \\ 4 \\ 1 \end{pmatrix}$$

and other vectors are given in the same styles, with **k** the unit vector in the $z$ direction.

The vector equation of a line is just like that in two dimensions. For example:

$$r = \begin{pmatrix} 3 \\ 4 \\ 1 \end{pmatrix} + \lambda \begin{pmatrix} 2 \\ 3 \\ 6 \end{pmatrix}$$

represents a line through the point with position vector $\begin{pmatrix} 3 \\ 4 \\ 1 \end{pmatrix}$, in the direction $\begin{pmatrix} 2 \\ 3 \\ 6 \end{pmatrix}$.

By contrast the cartesian form of a line in three dimensions is rather more complicated. The equation

$$\mathbf{r} = \begin{pmatrix} x \\ y \\ z \end{pmatrix} = \begin{pmatrix} 3 \\ 4 \\ 1 \end{pmatrix} + \lambda \begin{pmatrix} 2 \\ 3 \\ 6 \end{pmatrix}$$

contains three relationships, which are parametric equations for the line:

$$x = 3 + 2\lambda \qquad y = 4 + 3\lambda \qquad z = 1 + 6\lambda.$$

Making $\lambda$ the subject of each of these gives

$$\lambda = \frac{x-3}{2} \qquad \lambda = \frac{y-4}{3} \qquad \text{and} \qquad \lambda = \frac{z-1}{6}$$

which leads to

$$\frac{x-3}{2} = \frac{y-4}{3} = \frac{z-1}{6}.$$

This is the cartesian form of the equation of the line.

***Note***

The line's direction vector $\begin{pmatrix} 2 \\ 3 \\ 6 \end{pmatrix}$ can be read from the denominators of the three expressions in this equation, and a point through which it passes (3,4,1) from the three numerators.

The procedure may be generalised to write the equation of a straight line passing in direction **u** through a given point A with position vector **a**, as in figure 5.25.

$$\mathbf{a} = \begin{pmatrix} a_1 \\ a_2 \\ a_3 \end{pmatrix} \qquad \mathbf{u} = \begin{pmatrix} u_1 \\ u_2 \\ u_3 \end{pmatrix}$$

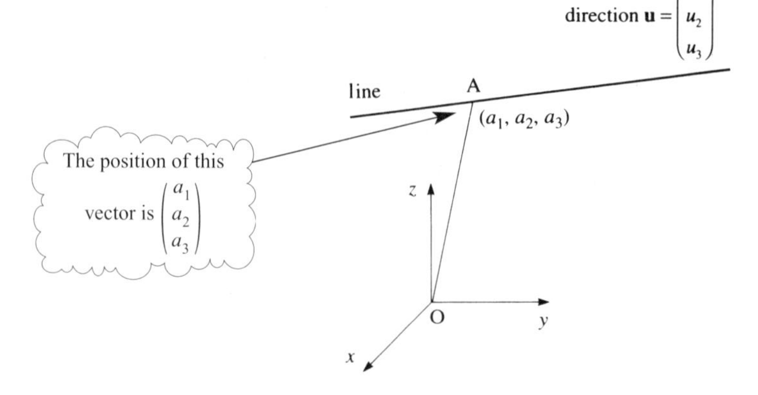


**Figure 5.25**

In vector form this is given by $\quad \mathbf{r} = \overrightarrow{OA} + \lambda\mathbf{u}$

which may be written as $\quad \mathbf{r} = \mathbf{a} + \lambda\mathbf{u}$

or in component form as

$$\mathbf{r} = \begin{pmatrix} x \\ y \\ z \end{pmatrix} = \begin{pmatrix} a_1 \\ a_2 \\ a_3 \end{pmatrix} + \lambda \begin{pmatrix} u_1 \\ u_2 \\ u_3 \end{pmatrix}.$$

This may then be written as the cartesian form of the equation:

$$\frac{x - a_1}{u_1} = \frac{y - a_2}{u_2} = \frac{z - a_3}{u_3}.$$

 The cartesian form involves two = symbols rather than one.

## Special cases of the cartesian form

In the general cartesian form of the equation of the straight line

$$\frac{x - a_1}{u_1} = \frac{y - a_2}{u_2} = \frac{z - a_3}{u_3}$$

the vector $\begin{pmatrix} u_1 \\ u_2 \\ u_3 \end{pmatrix}$ gives the direction of the line.

In this vector, at least one of $u_1$, $u_2$ and $u_3$ must be non-zero (otherwise the line would not be going anywhere and so would not be a line). However, there is no reason why more than one should be non-zero.

$\begin{pmatrix} 1 \\ 0 \\ 0 \end{pmatrix}$ and $\begin{pmatrix} 4 \\ 1 \\ 0 \end{pmatrix}$ are both valid directions.

In such cases, the equation of the line needs to be written differently, as in the following examples.

**EXAMPLE 5.14**

Find the cartesian form of the equation of the line through (7, 2, 3) in the direction $\begin{pmatrix} 0 \\ 5 \\ 2 \end{pmatrix}$.

**SOLUTION**

Substituting in the general form

$$\frac{x - a_1}{u_1} = \frac{y - a_2}{u_2} = \frac{z - a_2}{u_3}$$

gives

$$\frac{x - 7}{0} = \frac{y - 2}{5} = \frac{z - 3}{2}.$$

There is clearly a problem here since the first fraction involves division by 0. This difficulty is explained by the fact that for every point on the line $x - 7 = 0$, or $x = 7$. What was $\frac{x-7}{0}$ is now $\frac{0}{0}$; this is still undefined and so it is not equated to the other two expressions in the equation. Instead, the equation of the line is written

$$x = 7 \quad \text{and} \quad \frac{y - 2}{5} = \frac{z - 3}{2}.$$

**EXAMPLE 5.15**

Find the cartesian form of the equation of the line through (4, 2, 3) in the direction $\begin{pmatrix} 1 \\ 0 \\ 0 \end{pmatrix}$.

**SOLUTION**

Substituting in the general form

$$\frac{x - a_1}{u_1} = \frac{y - a_2}{u_2} = \frac{z - a_3}{u_3}$$

gives

$$\frac{x - 4}{1} = \frac{y - 2}{0} = \frac{z - 3}{0}.$$

The last two expressions tell you that $y = 2$ and $z = 3$.

The first part $\frac{x-4}{1}$ does not really give any further information: $x$ may take any value, and this is understood when the equation of the line is written as

$$y = 2 \quad \text{and} \quad z = 3.$$

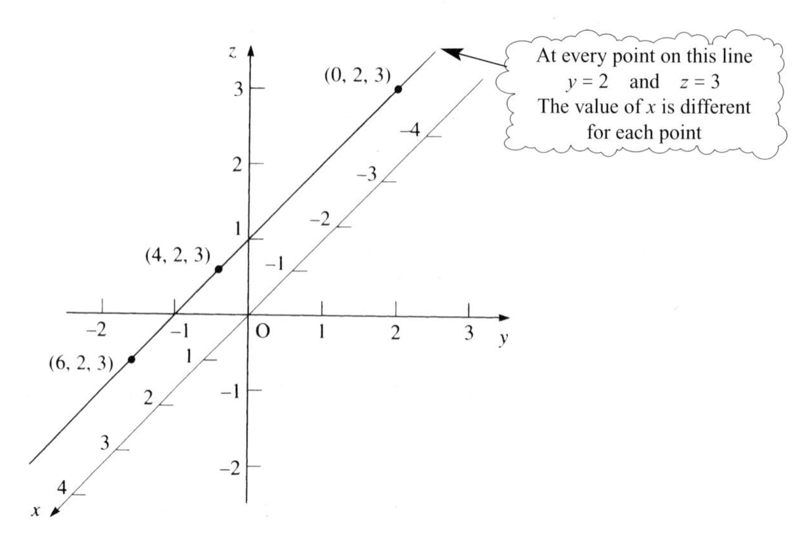


**Figure 5.26**

The vector forms of the equations of the lines given in the last two examples are

$$\mathbf{r} = \begin{pmatrix} x \\ y \\ z \end{pmatrix} = \begin{pmatrix} 7 \\ 2 \\ 3 \end{pmatrix} + \lambda \begin{pmatrix} 0 \\ 5 \\ 2 \end{pmatrix} \text{ and } \quad \mathbf{r} = \begin{pmatrix} x \\ y \\ z \end{pmatrix} = \begin{pmatrix} 4 \\ 2 \\ 3 \end{pmatrix} + \lambda \begin{pmatrix} 1 \\ 0 \\ 0 \end{pmatrix}.$$

These are considerably simpler than the equivalent cartesian forms. You will usually find it much easier to work with the equation of the line in vector form.

To convert from cartesian to vector form, you can reverse the procedure as in the following example. Usually, however, you would just write down the answer by looking at the numbers in the three numerators and denominators.

**EXAMPLE 5.16**

Write the equation of this line in vector form.

$$\frac{x-5}{2} = \frac{y+1}{1} = \frac{z+3}{6}$$

**SOLUTION**

$$\frac{x-5}{2} = \frac{y+1}{1} = \frac{z+3}{6} = \lambda$$

$$\frac{x-5}{2} = \lambda \quad \Rightarrow \quad x = 5 + 2\lambda$$

$$\frac{y+1}{1} = \lambda \quad \Rightarrow \quad y = -1 + \lambda$$

$$\frac{z+3}{6} = \lambda \quad \Rightarrow \quad z = -3 + 6\lambda$$

So

$$\mathbf{r} = \begin{pmatrix} x \\ y \\ z \end{pmatrix} = \begin{pmatrix} 5 + 2\lambda \\ -1 + \lambda \\ -3 + 6\lambda \end{pmatrix}$$

which is written

$$\mathbf{r} = \begin{pmatrix} 5 \\ -1 \\ -3 \end{pmatrix} + \lambda \begin{pmatrix} 2 \\ 1 \\ 6 \end{pmatrix}.$$

This line passes through (5, –1, –3) in the direction $\begin{pmatrix} 2 \\ 1 \\ 6 \end{pmatrix}$.

## The angle between two directions

When working in two dimensions you found the angle between two lines by using the scalar product. In the activity on page 153 you proved that this method can be extended into three dimensions, and its use is shown in the following example.

**EXAMPLE 5.17**

The points P, Q and R are (1, 0, –1), (2, 4, 1) and (3, 5, 6). Find ∠QPR.

**SOLUTION**

The angle between $\overrightarrow{PQ}$ and $\overrightarrow{PR}$ is given by $\theta$ in

$$\cos\theta = \frac{\overrightarrow{PQ} \cdot \overrightarrow{PR}}{|\overrightarrow{PQ}|\,|\overrightarrow{PR}|}$$

In this

$$\overrightarrow{PQ} = \begin{pmatrix} 2 \\ 4 \\ 1 \end{pmatrix} - \begin{pmatrix} 1 \\ 0 \\ -1 \end{pmatrix} = \begin{pmatrix} 1 \\ 4 \\ 2 \end{pmatrix} \qquad |\overrightarrow{PQ}| = \sqrt{1^2 + 4^2 + 2^2}$$
$$\sqrt{21}$$

Similarly

$$\overrightarrow{PR} = \begin{pmatrix} 3 \\ 5 \\ 6 \end{pmatrix} - \begin{pmatrix} 1 \\ 0 \\ -1 \end{pmatrix} = \begin{pmatrix} 2 \\ 5 \\ 7 \end{pmatrix} \qquad |\overrightarrow{PR}| = \sqrt{2^2 + 5^2 + 7^2}$$
$$= \sqrt{78}$$

Therefore

$$\overrightarrow{PQ} \cdot \overrightarrow{PR} = \begin{pmatrix} 1 \\ 4 \\ 2 \end{pmatrix} \cdot \begin{pmatrix} 2 \\ 5 \\ 7 \end{pmatrix}$$

$$= 1 \times 2 + 4 \times 5 + 2 \times 7$$
$$= 36$$

Substituting gives

$$\cos\theta = \frac{36}{\sqrt{21}\sqrt{78}}$$
$$\theta = 27.2°$$

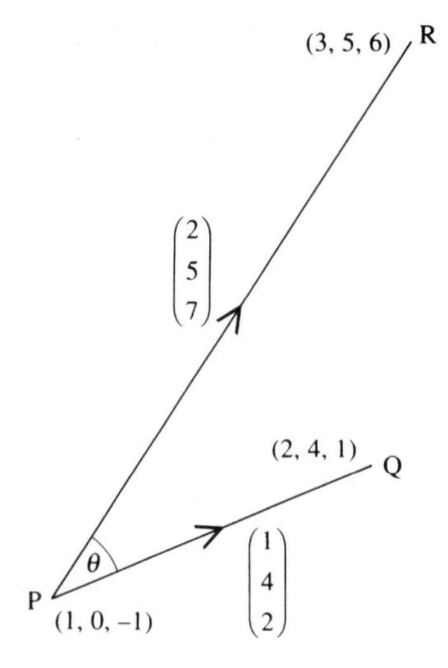


**Figure 5.27**

---

⚠ You must be careful to find the correct angle. To find ∠QPR (see figure 5.28), you need the scalar product $\overrightarrow{PQ}.\overrightarrow{PR}$. If you take $\overrightarrow{QP}.\overrightarrow{PR}$, you will obtain ∠Q'PR, which is 180° –∠QPR.

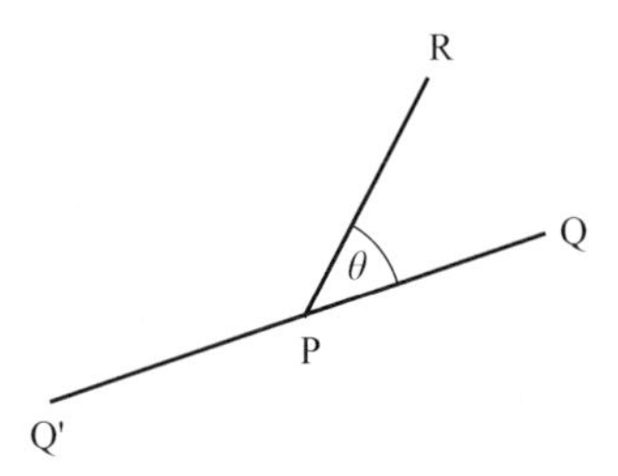


**Figure 5.28**

---

Even though two lines do not meet, it is still possible to specify the angle between them. The lines $l$ and $m$ shown in figure 5.29 do not meet; they are described as *skew*. The angle between them is that between their directions; it is shown in figure 5.29 as the angle $\theta$ between the lines $l$ and $m'$, where $m'$ is a translation of the line $m$ to a position where it does intersect the line $l$.

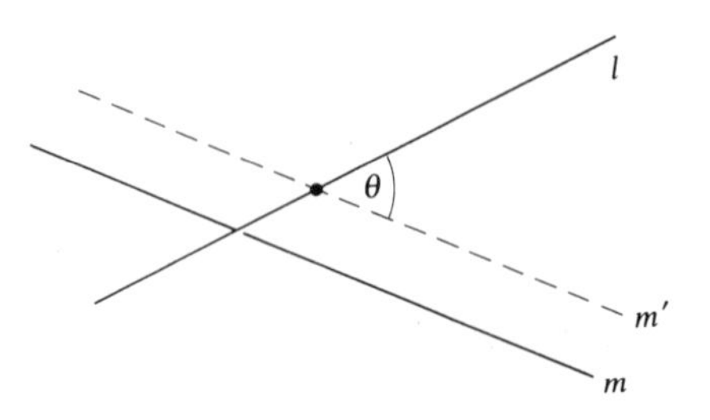


**Figure 5.29**

**EXAMPLE 5.18**

Find the angle between the lines

$$\mathbf{r} = \begin{pmatrix}1\\0\\4\end{pmatrix} + \lambda\begin{pmatrix}2\\-1\\-1\end{pmatrix} \quad \text{and} \quad \mathbf{r} + \begin{pmatrix}2\\-1\\3\end{pmatrix} + \mu\begin{pmatrix}3\\0\\1\end{pmatrix}.$$

**SOLUTION**

The angle between the lines is the angle between their directions $\begin{pmatrix}2\\-1\\-1\end{pmatrix}$ and $\begin{pmatrix}3\\0\\1\end{pmatrix}$.

Using $\cos\theta = \dfrac{\mathbf{a}.\mathbf{b}}{|\mathbf{a}|\,|\mathbf{b}|}$

$$\cos\theta = \frac{2\times 3 + (-1)\times 0 + (-1)\times 1}{\sqrt{2^2 + (-1)^2 + (-1)^2}\ \sqrt{3^2 + 0^2 + 1^2}}$$

$$\cos\theta = \frac{5}{\sqrt{6}\sqrt{10}}$$

$$\theta = 49.8°.$$

**EXERCISE 5E**

**1** Find the equations of the following lines in vector form.

**(i)** through (2, 4, –1) in the direction $\begin{pmatrix}3\\6\\4\end{pmatrix}$

**(ii)** through (1, 0, –1) in the direction $\begin{pmatrix}1\\0\\0\end{pmatrix}$

**(iii)** through (1, 0, 4) and (6, 3, –2)

**(iv)** through (0, 0, 1) and (2, 1, 4)

**(v)** through (1, 2, 3) and (–2, –4, –6)

**2** Write the equations of the following lines in cartesian form.

**(i)** $\mathbf{r} = \begin{pmatrix}2\\4\\-1\end{pmatrix} + \lambda\begin{pmatrix}3\\6\\4\end{pmatrix}$

**(ii)** $\mathbf{r} = \begin{pmatrix}1\\0\\-1\end{pmatrix} + \lambda\begin{pmatrix}1\\3\\4\end{pmatrix}$

**(iii)** $\mathbf{r} = \begin{pmatrix}3\\0\\4\end{pmatrix} + \lambda\begin{pmatrix}1\\0\\2\end{pmatrix}$

**(iv)** $\mathbf{r} = \begin{pmatrix}0\\4\\1\end{pmatrix} + \lambda\begin{pmatrix}2\\0\\4\end{pmatrix}$

**(v)** $\mathbf{r} = \begin{pmatrix}-2\\-7\\3\end{pmatrix} + \lambda\begin{pmatrix}0\\1\\0\end{pmatrix}$

**3** Write the equations of the following lines in vector form.

**(i)** $\dfrac{x-3}{5} = \dfrac{y+2}{3} = \dfrac{z-1}{4}$

**(ii)** $\dfrac{x+6}{6} = \dfrac{y}{2} = \dfrac{z+4}{3}$

**(iii)** $x = \dfrac{y}{2} = \dfrac{z+1}{3}$

**(iv)** $x = y = z$

**(v)** $x = 2$ and $y = z$

**4** Find the angles between these pairs of vectors.

**(i)** $\begin{pmatrix}2\\1\\3\end{pmatrix}$ and $\begin{pmatrix}2\\-1\\4\end{pmatrix}$ **(ii)** $\begin{pmatrix}1\\-1\\0\end{pmatrix}$ and $\begin{pmatrix}3\\1\\5\end{pmatrix}$

**(iii)** $3\mathbf{i} + 2\mathbf{j} - 2\mathbf{k}$ and $-4\mathbf{i} - \mathbf{j} + 3\mathbf{k}$

**5** Find the angles between these pairs of lines.

**(i)** $\mathbf{r} = \begin{pmatrix}2\\1\\3\end{pmatrix} + \lambda\begin{pmatrix}1\\4\\0\end{pmatrix}$ and $\mathbf{r} = \begin{pmatrix}6\\10\\4\end{pmatrix} + \lambda\begin{pmatrix}2\\1\\1\end{pmatrix}$

**(ii)** $\mathbf{r} = \lambda\begin{pmatrix}4\\1\\4\end{pmatrix}$ and $\mathbf{r} = \begin{pmatrix}7\\0\\-3\end{pmatrix} + \lambda\begin{pmatrix}1\\2\\-1\end{pmatrix}$

**(iii)** $\dfrac{x-4}{3} = \dfrac{y-2}{7} = \dfrac{z+1}{-4}$ and $\dfrac{x-5}{2} = \dfrac{y-1}{8} = \dfrac{z}{-5}$

**6** The room illustrated in the diagram has rectangular walls, floor and ceiling. A string has been stretched in a straight line between the corners A and G.

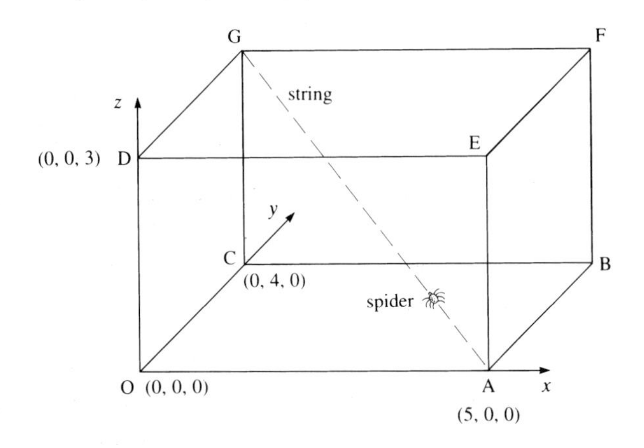


The corner O is taken as the origin. A is (5, 0, 0), C is (0, 4, 0) and D is (0, 0, 3), where the lengths are in metres.

**(i)** Write down the co-ordinates of G.

**(ii)** Find the vector $\overrightarrow{AG}$ and the length of the string $|\overrightarrow{AG}|$.

**(iii)** Write down the equation of the line AG in vector form.

A spider walks up the string, starting from A.

**(iv)** Find the position vector of the spider when it is at Q, one quarter of the way from A to G, and find the angle OQG.

**(v)** Show that when the spider is 1.5 m above the floor it is at its closest point to O, and find how far it is then from O.

[MEI]

**7** The diagram shows an extension to a house. Its base and walls are rectangular and the end of its roof, EPF, is sloping, as illustrated.

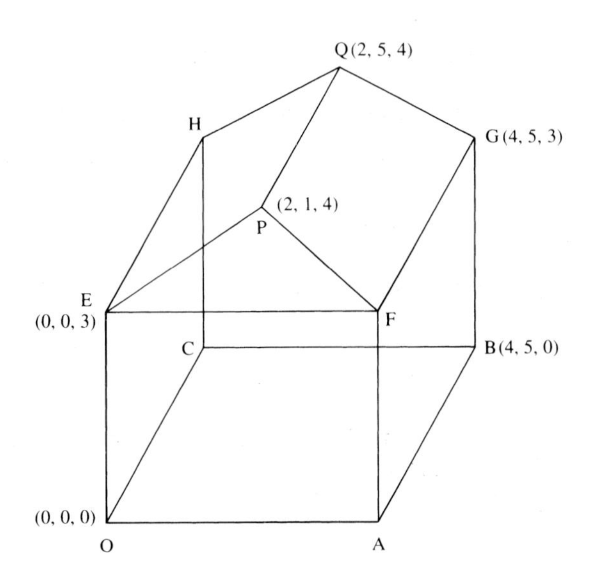


**(i)** Write down the co-ordinates of A and F.

**(ii)** Find, using vector methods, the angles FPQ and EPF.

The owner decorates the room with two streamers which are pulled taut. One goes from O to G, the other from A to H. She says that they touch each other and that they are perpendicular to each other.

**(iii)** Is she right?

**8** The drawing shows an ordinary music stand, which consists of a rectangle DEFG with a vertical support OA. Relative to axes through the origin O, which is on the floor, the co-ordinates of various points are given (with dimensions in metres) as:

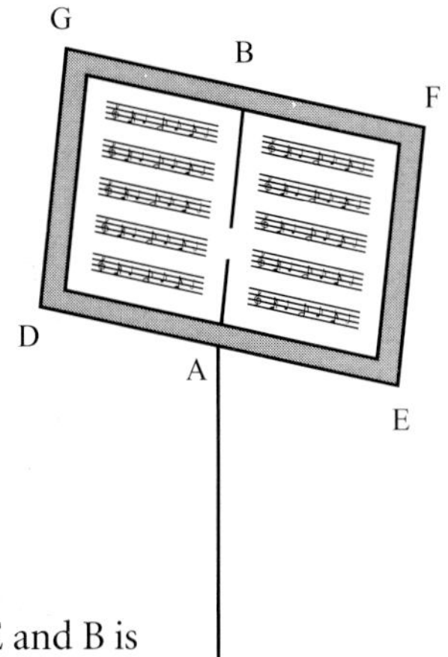


A is (0, 0, 1)

D is (–0.25, 0, 1)

F is (0.25, 0.15, 1.3)

DE and GF are horizontal, A is the mid-point of DE and B is the mid-point of GF.

C is on AB so that AC = $\frac{1}{3}$AB.

**(i)** Write down the vector $\overrightarrow{AD}$ and show that $\overrightarrow{EF}$ is $\begin{pmatrix} 0 \\ 0.15 \\ 0.3 \end{pmatrix}$.

**(ii)** Calculate the co-ordinates of C.

**(iii)** Find the equations of the lines DE and EF in vector form.

[MEI, part]

## Planes

? Which balances better, a three-legged stool or a four-legged stool? Why? What information do you need to specify a particular plane?

There are various ways of finding the equation of a plane. Two of these are given in this book. Your choice of which one to use will depend on the information you are given.

### THE EQUATION OF A PLANE, GIVEN THREE POINTS ON IT

There are several methods used to find the equation of a plane through three given points. The shortest method involves the use of vector product which is beyond the scope of this book but is covered in *Pure Mathematics 4*. The method given here (pages 162–164) develops the same ideas as were used for the equation of a line. It will help you to understand the extra concepts involved, but it is not a requirement of the MEI Pure Mathematics 3 subject criteria.

To find the vector form of the equation of the plane through the points A, B and C (with position vectors $\overrightarrow{OA} = \mathbf{a}$, $\overrightarrow{OB} = \mathbf{b}$, $\overrightarrow{OC} = \mathbf{c}$), think of starting at the origin, travelling along OA to join the plane at A, and then any distance in each of the directions $\overrightarrow{AB}$ and $\overrightarrow{AC}$ to reach a general point R with position vector $\mathbf{r}$, where

$$\mathbf{r} = \overrightarrow{OA} + \lambda\overrightarrow{AB} + \mu\overrightarrow{AC}.$$

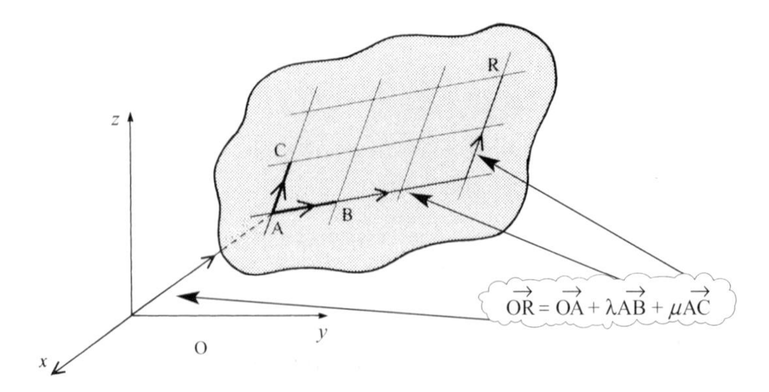


**Figure 5.30**

This is a vector form of the equation of the plane. Since $\overrightarrow{OA} = \mathbf{a}$, $\overrightarrow{AB} = \mathbf{b} - \mathbf{a}$ and $\overrightarrow{AC} = \mathbf{c} - \mathbf{a}$, it may also be written as

$$\mathbf{r} = \mathbf{a} + \lambda(\mathbf{b} - \mathbf{a}) + \mu(\mathbf{c} - \mathbf{a}).$$





**EXAMPLE 5.19**

Find the equation of the plane through A(4, 2, 0), B(3, 1, 1) and C(4, –1, 1).

**SOLUTION**

$$\overrightarrow{OA} = \begin{pmatrix} 4 \\ 2 \\ 0 \end{pmatrix}$$

$$\overrightarrow{AB} = \overrightarrow{OB} - \overrightarrow{OA} = \begin{pmatrix} 3 \\ 1 \\ 1 \end{pmatrix} - \begin{pmatrix} 4 \\ 2 \\ 0 \end{pmatrix} = \begin{pmatrix} -1 \\ -1 \\ 1 \end{pmatrix}$$

$$\overrightarrow{AC} = \overrightarrow{OC} - \overrightarrow{OA} = \begin{pmatrix} 4 \\ -1 \\ 1 \end{pmatrix} - \begin{pmatrix} 4 \\ 2 \\ 0 \end{pmatrix} = \begin{pmatrix} 0 \\ -3 \\ 1 \end{pmatrix}$$

So the equation $\mathbf{r} = \overrightarrow{OA} + \lambda\overrightarrow{AB} + \mu\overrightarrow{AC}$ becomes

$$\mathbf{r} = \begin{pmatrix} 4 \\ 2 \\ 0 \end{pmatrix} + \lambda\begin{pmatrix} -1 \\ -1 \\ 1 \end{pmatrix} + \mu\begin{pmatrix} 0 \\ -3 \\ 1 \end{pmatrix}.$$

This is the vector form of the equation, written using components.

**CARTESIAN FORM**

You can convert this equation into cartesian form by writing it as

$$\begin{pmatrix} x \\ y \\ z \end{pmatrix} = \begin{pmatrix} 4 \\ 2 \\ 0 \end{pmatrix} + \lambda\begin{pmatrix} -1 \\ -1 \\ 1 \end{pmatrix} + \mu\begin{pmatrix} 0 \\ -3 \\ 1 \end{pmatrix}$$

and eliminating $\lambda$ and $\mu$. The three equations contained in this vector equation may be simplified to give

$$\lambda = -x + 4 \qquad ①$$

$$\lambda + 3\mu = -y + 2 \qquad ②$$

$$\lambda + \mu = z. \qquad ③$$

Substituting ① into ② gives

$$-x + 4 + 3\mu = -y + 2$$

$$3\mu = x - y - 2$$

$$\mu = \tfrac{1}{3}(x - y - 2).$$

Substituting this and ① into ③ gives

$$-x + 4 + \tfrac{1}{3}(x - y - 2) = z$$

$$-3x + 12 + x - y - 2 = 3z$$

$$2x + y + 3z = 10$$

and this is the cartesian equation of the plane through A, B and C.

*Note*

In contrast to the equation of a line, the equation of a plane is more neatly expressed in cartesian form. The general cartesian equation of a plane is often written as either

$$ax + by + cz + d = 0 \quad \text{or} \quad n_1x + n_2y + n_3z + d = 0.$$

## The direction perpendicular to a plane

? Lay a sheet of paper on a flat horizontal table and mark several straight lines on it. Now take a pencil and stand it upright on the sheet of paper (see figure 5.31).

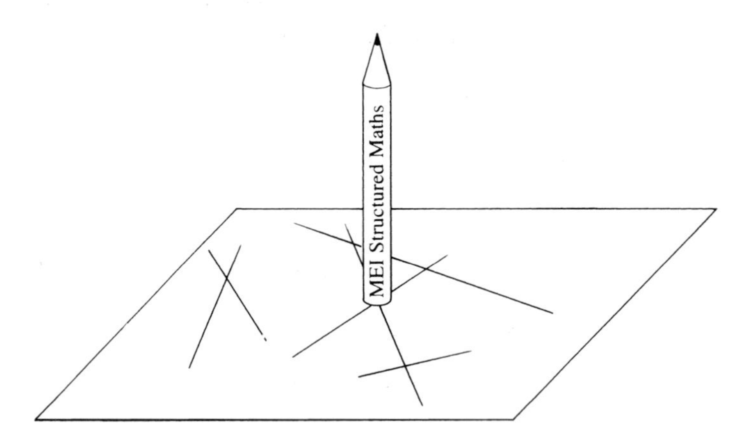


**Figure 5.31**

**(i)** What angle does the pencil make with any individual line?

**(ii)** Would it make any difference if the table were tilted at an angle (apart from the fact that you could no longer balance the pencil)?

The above discussion shows you that there is a direction (that of the pencil) which is at right angles to every straight line in the plane. A line in that direction is said to be perpendicular to the plane.

This allows you to find a different vector form of the equation of a plane which you use when you know the position vector **a** of one point A on the plane and the direction $\mathbf{n} = n_1\mathbf{i} + n_2\mathbf{j} + n_3\mathbf{k}$ perpendicular to the plane.

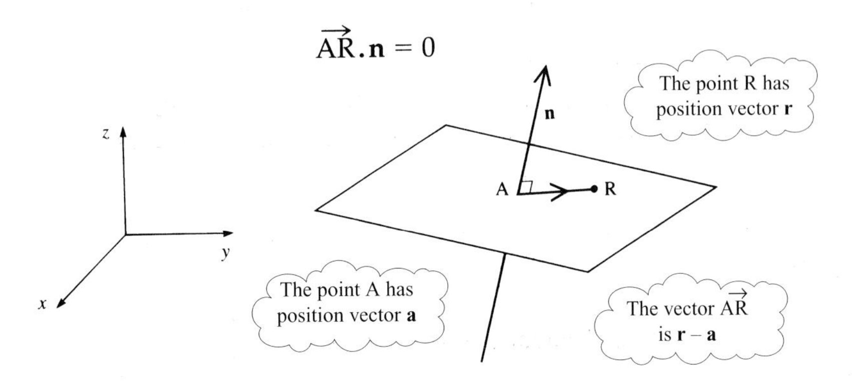


**Figure 5.32**

What you want to find is an expression for the position vector **r** of a general point R on the plane (figure 5.32). Since AR is a line in the plane, it follows that AR is at right angles to the direction **n**.

$$\overrightarrow{AR}.\mathbf{n} = 0$$

The vector $\overrightarrow{AR}$ is given by

$$\overrightarrow{AR} = \mathbf{r} - \mathbf{a}$$

and so $(\mathbf{r} - \mathbf{a}).\mathbf{n} = 0.$

This can also be written as

$$\mathbf{r}.\mathbf{n} - \mathbf{a}.\mathbf{n} = 0$$

or $$\begin{pmatrix} x \\ y \\ z \end{pmatrix} . \begin{pmatrix} n_1 \\ n_2 \\ n_3 \end{pmatrix} - \mathbf{a}.\mathbf{n} = 0$$

$$\Rightarrow n_1x + n_2y + n_3z + d = 0$$

where $d = -\mathbf{a}.\mathbf{n}.$

Notice that $d$ is a constant scalar.

**EXAMPLE 5.20**

Write down the equation of the plane through the point (2, 1, 3) given that the vector $\begin{pmatrix}4\\5\\6\end{pmatrix}$ is perpendicular to the plane.

**SOLUTION**

In this case, the position vector **a** of the point (2, 1, 3) is given by $\mathbf{a} = \begin{pmatrix}2\\1\\3\end{pmatrix}$.

The vector perpendicular to the plane is

$$\mathbf{n} = \begin{pmatrix}n_1\\n_2\\n_3\end{pmatrix} = \begin{pmatrix}4\\5\\6\end{pmatrix}.$$

The equation of the plane is

$$n_1x + n_2y + n_3z - \mathbf{a.n} = 0$$

$$4x + 5y + 6z - (2 \times 4 + 1 \times 5 + 3 \times 6) = 0$$

$$4x + 5y + 6z - 31 = 0.$$

Look carefully at the equation of the plane in Example 5.20. You can see at once that the vector $\begin{pmatrix}4\\5\\6\end{pmatrix}$, formed from the coefficients of $x$, $y$ and $z$, is perpendicular to the plane.

The vector $\begin{pmatrix}n_1\\n_2\\n_3\end{pmatrix}$ is perpendicular to all planes of the form

$$n_1x + n_2y + n_3z + d = 0$$

whatever the value of $d$ (figure 5.33). Consequently, all planes of that form are parallel; the coefficients of $x$, $y$ and $z$ determine the direction of the plane, the value of $d$ its location.

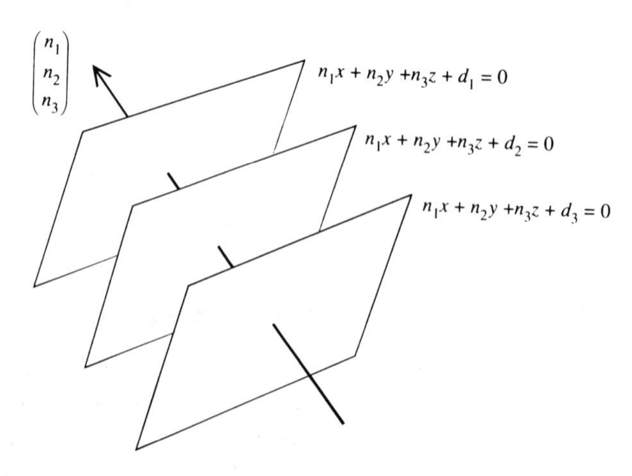


**Figure 5.33**

### The intersection of a line and a plane

The point of intersection of a line and a plane is found by following the procedure in the next example.

**EXAMPLE 5.21**

Find the point of intersection of the line

$$\mathbf{r} = \begin{pmatrix} 2 \\ 3 \\ 4 \end{pmatrix} + \lambda \begin{pmatrix} 1 \\ 2 \\ -1 \end{pmatrix}$$

with the plane $5x + y - z = 1$.

**SOLUTION**

The line is

$$\mathbf{r} = \begin{pmatrix} x \\ y \\ z \end{pmatrix} = \begin{pmatrix} 2 \\ 3 \\ 4 \end{pmatrix} + \lambda \begin{pmatrix} 1 \\ 2 \\ -1 \end{pmatrix}$$

and so for any point on the line

$$x = 2 + \lambda \qquad y = 3 + 2\lambda \qquad \text{and} \qquad z = 4 - \lambda.$$

Substituting these into the equation of the plane $5x + y - z = 1$ gives

$$5(2 + \lambda) + (3 + 2\lambda) - (4 - \lambda) = 1$$

$$8\lambda = -8$$

$$\lambda = -1.$$

Substituting $\lambda = -1$ in the equation of the line gives

$$\mathbf{r} = \begin{pmatrix} x \\ y \\ z \end{pmatrix} = \begin{pmatrix} 1 \\ 1 \\ 5 \end{pmatrix}$$

so the point of intersection is (1, 1, 5).

As a check, substitute (1, 1, 5) into the equation of the plane:

$$5x + y - z = 1$$

$$5 + 1 - 5 = 1$$

$$1 = 1 \qquad \text{as required.}$$

### The distance of a point from a plane

The shortest distance of a point, A, from a plane is the distance AP, where P is the point where the line through A perpendicular to the plane intersects the plane (figure 5.34, overleaf). This is usually just called the distance of the point from the plane. The process of finding this distance is shown in the next example.

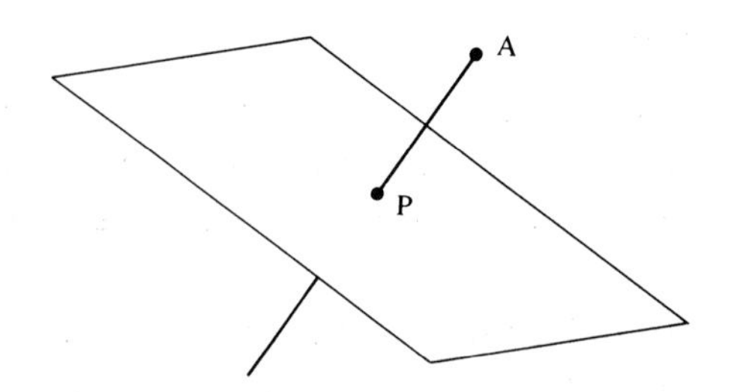


**Figure 5.34**

**EXAMPLE 5.22**

A is the point (7, 5, 3) and the plane $\pi$ has the equation $3x + 2y + z = 6$. Find

**(i)** the equation of the line through A perpendicular to the plane $\pi$;

**(ii)** the point of intersection, P, of this line with the plane;

**(iii)** the distance AP.

**SOLUTION**

**(i)** The direction perpendicular to the plane $3x + 2y + z = 6$ is $\begin{pmatrix} 3 \\ 2 \\ 1 \end{pmatrix}$ so the line through (7, 5, 3) perpendicular to the plane is given by

$$\mathbf{r} = \begin{pmatrix} 7 \\ 5 \\ 3 \end{pmatrix} + \lambda \begin{pmatrix} 3 \\ 2 \\ 1 \end{pmatrix}.$$

**(ii)** For any point on the line

$$x = 7 + 3\lambda \qquad y = 5 + 2\lambda \qquad \text{and} \qquad z = 3 + \lambda.$$

Substituting these expressions into the equation of the plane $3x + 2y + z = 6$ gives

$$3(7 + 3\lambda) + 2(5 + 2\lambda) + (3 + \lambda) = 6$$

$$14\lambda = -28$$

$$\lambda = -2.$$

So the point P has co-ordinates (1, 1, 1).

**(iii)** The vector $\overrightarrow{AP}$ is given by

$$\begin{pmatrix} 1 \\ 1 \\ 1 \end{pmatrix} - \begin{pmatrix} 7 \\ 5 \\ 3 \end{pmatrix} = \begin{pmatrix} -6 \\ -4 \\ -2 \end{pmatrix}$$

and so the length AP is

$$\sqrt{(-6)^2 + (-4)^2 + (-2)^2} = \sqrt{56}.$$

*Note*

In practice, you would usually not follow the procedure in Example 5.22 because there is a well-known formula for the distance of a point from a plane. You are invited to derive this in the following activity.

**ACTIVITY**

Generalise the work in Example 5.22 to show that the distance of the point $(\alpha, \beta, \gamma)$ from the line $n_1x + n_2y + n_3z + d = 0$ is given by

$$\frac{|n_1\alpha + n_2\beta + n_3\gamma + d|}{\sqrt{n_1^2 + n_2^2 + n_3^2}}$$

**EXERCISE 5F**

**1** The points A, B and C have co-ordinates (0, 1, 1), (–2, –1, –5) and (1, –1, 0).

**(i)** Find the vectors $\overrightarrow{AB}$ and $\overrightarrow{AC}$.

**(ii)** Find the equation of the plane ABC in the form

$$\mathbf{r} = \overrightarrow{OA} + \lambda\overrightarrow{AB} + \mu\overrightarrow{AC}.$$

**(iii)** Verify that A, B and C lie on the plane $5x + 4y - 3z = 1$.

**(iv)** Show that

$$\overrightarrow{AB}.\begin{pmatrix}5\\4\\-3\end{pmatrix} = \overrightarrow{BC}.\begin{pmatrix}5\\4\\-3\end{pmatrix} = 0$$

and explain the significance of these results.

**2** The points L, M and N have co-ordinates (0, –1, 2), (2, 1, 0) and (5, 1, 1).

**(i)** Write down the vectors $\overrightarrow{LM}$ and $\overrightarrow{LN}$.

**(ii)** Show that

$$\overrightarrow{LM}.\begin{pmatrix}1\\-4\\-3\end{pmatrix} = \overrightarrow{LN}.\begin{pmatrix}1\\-4\\-3\end{pmatrix} = 0.$$

**(iii)** Find the equation of the plane LMN.

**3** The points A, B and C have co-ordinates (3, 0, 0), (3, 1, 2) and (3, 4, –2).

**(i)** Show that the equation of the plane ABC may be written as

$$\mathbf{r} = \begin{pmatrix}3\\0\\0\end{pmatrix} + \lambda\begin{pmatrix}0\\1\\2\end{pmatrix} + \mu\begin{pmatrix}0\\2\\-1\end{pmatrix}.$$

**(ii)** Show that the equation of the plane may also be written in the form $x = 3$.

**(iii)** Describe this plane.

**4 (i)** Show that the points A(1, 1, 1), B(3, 0, 0) and C(2, 0, 2) all lie in the plane $2x + 3y + z = 6$.

**(ii)** Show that

$$\overrightarrow{AB}.\begin{pmatrix}2\\3\\1\end{pmatrix} = \overrightarrow{AC}.\begin{pmatrix}2\\3\\1\end{pmatrix} = 0.$$

**(iii)** The point D has co-ordinates (7, 6, 2). D lies on a line perpendicular to the plane through one of the points A, B or C. Through which of these points does the line pass?

**5** The lines $l$, $\mathbf{r} = \begin{pmatrix}2\\1\\0\end{pmatrix} + \lambda\begin{pmatrix}1\\1\\1\end{pmatrix}$, and $m$, $\mathbf{r} = \begin{pmatrix}4\\0\\2\end{pmatrix} + \mu\begin{pmatrix}1\\0\\1\end{pmatrix}$, lie in the same plane $\pi$.

**(i)** Find the co-ordinates of any two points on each of the lines.

**(ii)** Show that all the four points you found in part (i) lie on the plane $x - z = 2$.

**(iii)** Explain why you now have more than sufficient evidence to show that the plane $\pi$ has equation $x - z = 2$.

**(iv)** Find the co-ordinates of the point where the lines $l$ and $m$ intersect.

**6** Find the points of intersection of the following planes and lines.

**(i)** $x + 2y + 3z = 11$ and $\mathbf{r} = \begin{pmatrix}1\\2\\4\end{pmatrix} + \lambda\begin{pmatrix}1\\1\\1\end{pmatrix}$

**(ii)** $2x + 3y - 4z = 1$ and $\frac{x+2}{3} = \frac{y+3}{4} = \frac{z+4}{5}$

**(iii)** $3x - 2y - z = 14$ and $\mathbf{r} = \begin{pmatrix}8\\4\\2\end{pmatrix} + \lambda\begin{pmatrix}1\\2\\1\end{pmatrix}$

**(iv)** $x + y + z = 0$ and $\mathbf{r} = \lambda\begin{pmatrix}1\\1\\2\end{pmatrix}$

**(v)** $5x - 4y - 7z = 49$ and $\frac{x-3}{2} = \frac{y+1}{5} = \frac{z-2}{-3}$

**7** In each of the following examples you are given a point A and a plane $\pi$. Find:

**(a)** the equation of the line through A perpendicular to $\pi$;

**(b)** the point of intersection, P, of this line with $\pi$;

**(c)** the distance AP.

**(i)** A is (2, 2, 3); $\pi$ is $x - y + 2z = 0$

**(ii)** A is (2, 3, 0); $\pi$ is $2x + 5y + 3z = 0$

**(iii)** A is (3, 1, 3); $\pi$ is $x = 0$

**(iv)** A is (2, 1, 0); $\pi$ is $3x - 4y + z = 2$

**(v)** A is (0, 0, 0); $\pi$ is $x + y + z = 6$

**8** The points U and V have co-ordinates (4, 0, 7) and (6, 4, 13). The line UV is perpendicular to a plane and the point U lies on the plane.

**(i)** Find the equation of the plane in cartesian form.

**(ii)** The point W has co-ordinates (–1, 10, 2). Show that $WV^2 = WU^2 + UV^2$.

**(iii)** What information does this give you about the position of W? Confirm this information by a different method.

**9** **(i)** Find the equation of the line through (13, 5, 0) parallel to the line

$$\mathbf{r} = \begin{pmatrix} 2 \\ -1 \\ 4 \end{pmatrix} + \lambda \begin{pmatrix} 3 \\ 1 \\ -2 \end{pmatrix}.$$

**(ii)** Where does this line meet the plane $3x + y - 2z = 2$?

**(iii)** How far is the point of intersection from (13, 5, 0)?

**10** A is the point (1, 2, 0), B(0, 4, 1) and C(9, –2, 1).

**(i)** Show that A, B and C lie on the plane $2x + 3y - 4z = 8$.

**(ii)** Write down the vectors $\overrightarrow{AB}$ and $\overrightarrow{AC}$ and verify that they are at right angles to $\begin{pmatrix} 2 \\ 3 \\ -4 \end{pmatrix}$.

**(iii)** Find the angle BAC.

**(iv)** Find the area of triangle ABC (using area = $\frac{1}{2}bc\sin A$).

**11** P is the point (2, –1, 3), Q is (5, –5, 3) and R is (7, 2, –3). Find

**(i)** the lengths of **(a)** PQ **(b)** QR;

**(ii)** the angle PQR;

**(iii)** the area of triangle PQR;

**(iv)** the point S such that PQRS is a parallelogram.

**12** P is the point (2, 2, 4), Q(0, 6, 8), X(–2, –2, –3) and Y(2, 6, 9).

**(i)** Write in vector form the equations of the lines PQ and XY.

**(ii)** Verify that the equation of the plane PQX is $2x + 5y - 4z = -2$.

**(iii)** Does the point Y lie on the plane PQX?

**(iv)** Does any point on PQ lie on XY? (That is, do the lines intersect?)

**13** **(i)** Find, in vector or cartesian form, the equation of a line passing through the two points A(4, 1, 3) and B(6, 4, 8).

**(ii)** Find the co-ordinates of the point P where the line which you have found in part **(i)** meets the plane $x + 2y - z + 3 = 0$.

A line is drawn through A perpendicular to the plane.

**(iii)** Find the coordinates of the point Q where this line cuts the plane and also the co-ordinates of the point $A_1$, the mirror image of the point A in the plane.

**(iv)** Use scalar products to calculate the angles PAQ and $PA_1Q$.

[MEI]

**14** You are given the four points O(0, 0, 0), A(5, –12, 16), B(8, 3, 19) and C(–23, –80, 12).

**(i)** Show that the three points A, B and C all lie in the plane with equation $2x - y + 3z = 70$.

**(ii)** Write down a vector which is normal to this plane.

**(iii)** The line from the origin O perpendicular to this plane meets the plane at D. Find the co-ordinates of D.

**(iv)** Write down the equations of the two lines OA and AB in vector form.

**(v)** Hence find the angle OAB, correct to the nearest degree.

[MEI]

**15 (i)** Write down in vector or cartesian form the equation of the line joining A(8, 0, –4) to B(12, 2, –6).

**(ii)** This line meets the plane $2x + y - z = 2$ at C. Find the co-ordinates of C.

**(iii)** Find the length of the line joining C to B.

**(iv)** Find the ratio in which the point A divides CB.

**(v)** Find the angle AOB where O is the origin.

[MEI]

**16**

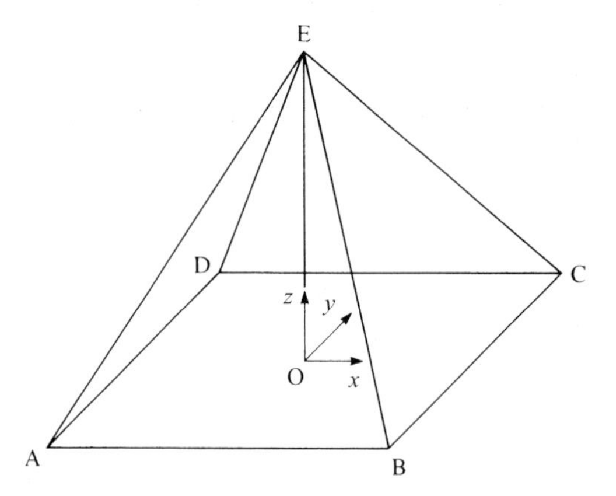


An explorer comes across a hollow pyramid with a square base of side 100 m and with height 100 m. Taking the origin to be the middle point of the base and 1 unit to be 1 m:

**(i)** write down the co-ordinates of the vertices of the pyramid;

**(ii)** show that the face BCE has equation $2x + z = 100$ and write down the equations of the other three sloping faces.

The explorer finds that inside the pyramid a rope is hanging from vertex E, and begins to climb it.

**(iii)** When he has climbed 20 m, he shines his torch directly on to the face BCE. Find the equation of the line of the torch beam, in vector form, and hence find how far the explorer is from the face.

**(iv)** When the explorer has climbed to a height $h$ metres, he is the same distance from the ground as he is from each of the sloping faces. Show that

$$h = \frac{100}{1 + \sqrt{5}}.$$

**17** In bad weather, the roof of a barn begins to sag. It is decided to support it as shown in the diagram.

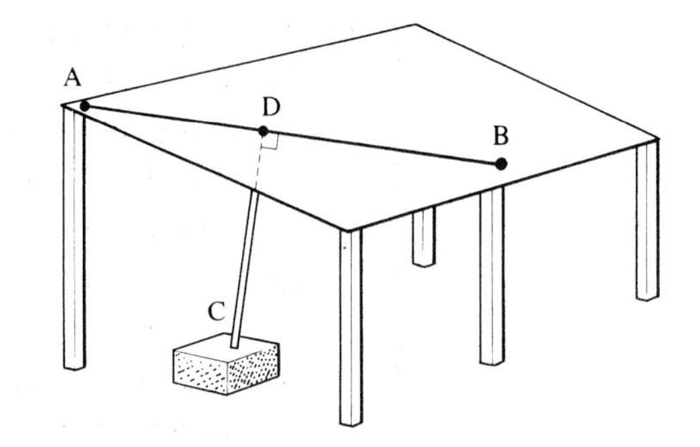


When the roof is supported, ADB is a straight line. Two points on the roof are A(2, 0, 15) and B(14, 9, 9) relative to an arbitrary origin.

**(i)** Find the equation of the line AB in vector form.

The support CD, resting on concrete blocks at C, is perpendicular to the line AB. C is the point (3, – 1, 1).

**(ii)** Write down the vector $\overrightarrow{CP}$, where P is a general point on the line AB. Hence, using a scalar product, find the co-ordinates of D on AB such that CD is perpendicular to AB.

**(iii)** Calculate the length of the support CD.

**(iv)** Calculate the ratio AD:DB.

[MEI]

**18** A pyramid in the shape of a tetrahedron has base ABC and vertex P as shown in the diagram. The vertices A, B, C, P have position vectors

$\mathbf{a} = -4\mathbf{j} + 2\mathbf{k}$,

$\mathbf{b} = 2\mathbf{i} + 4\mathbf{k}$,

$\mathbf{c} = -5\mathbf{i} - 2\mathbf{j} + 6\mathbf{k}$,

$\mathbf{p} = 3\mathbf{i} - 8\mathbf{j} + 12\mathbf{k}$

respectively.

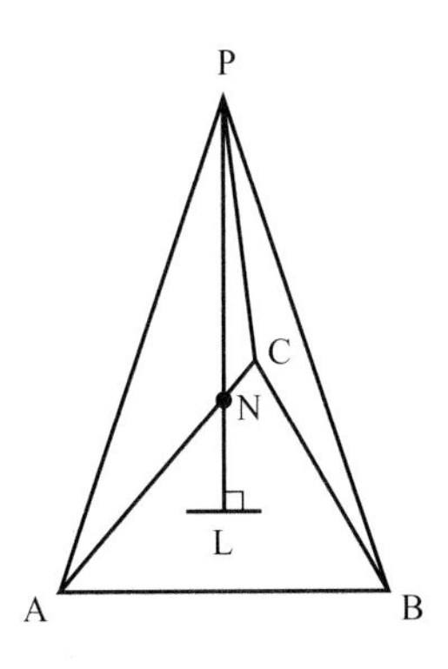


The equation of the plane of the base is

$$\mathbf{r}.\begin{pmatrix} 2 \\ -3 \\ 4 \end{pmatrix} = 20.$$

**(i)** Write down a vector which is normal to the base ABC.

The line through P, perpendicular to the base, cuts the base at L.

**(ii)** Find the equation of the line PL in vector form and use it to find the co-ordinates of L.

**(iii)** Find the co-ordinates of the point N on LP, such that $\overrightarrow{LN} = \frac{1}{4}\overrightarrow{LP}$.

**(iv)** Find the angle between PA and PL.

[MEI]

**19** The diagram shows an arrow embedded in a target. The line of the arrow passes through the point A(2, 3, 5) and has direction vector $3\mathbf{i} + \mathbf{j} - 2\mathbf{k}$. The arrow intersects the target at the point B. The plane of the target has equation $x + 2y - 3z = 4$. The units are metres.

**(i)** Write down the vector equation of the line of the arrow in the form

$$\mathbf{r} = \mathbf{p} + \lambda\mathbf{q}.$$

**(ii)** Find the value of $\lambda$ which corresponds to B. Hence write down the co-ordinates of B.

**(iii)** The point C is where the line of the arrow meets the ground, which is the plane $z = 0$. Find the co-ordinates of C.

**(iv)** The tip, T, of the arrow is one-third of the way from B to C. Find the co-ordinates of T and the length of BT.

**(v)** Write down a normal vector to the plane of the target. Find the acute angle between the arrow and this normal.

A
B
T
C

[MEI]

**20** The position vectors of three points A, B, C on a plane ski-slope are

$$\mathbf{a} = 4\mathbf{i} + 2\mathbf{j} - \mathbf{k}, \quad \mathbf{b} = -2\mathbf{i} + 26\mathbf{j} + 11\mathbf{k}, \quad \mathbf{c} = 16\mathbf{i} + 17\mathbf{j} + 2\mathbf{k},$$

where the units are metres.

**(i)** Show that the vector $2\mathbf{i} - 3\mathbf{j} + 7\mathbf{k}$ is perpendicular to $\overrightarrow{AB}$ and also perpendicular to $\overrightarrow{AC}$. Hence find the equation of the plane of the ski-slope.

The track for an overhead railway lies along DEF, where D and E have position vectors $\mathbf{d} = 130\mathbf{i} - 40\mathbf{j} + 20\mathbf{k}$ and $\mathbf{e} = 90\mathbf{i} - 20\mathbf{j} + 15\mathbf{k}$, and F is a point on the ski-slope.

**(ii)** Find the equation of the straight line DE.

**(iii)** Find the position vector of the point F.

**(iv)** Find the length of the track DF.

[MEI]

**21** A plane $\pi$ has equation $ax + by + z = d$.

**(i)** Write down, in terms of $a$ and $b$, a vector which is perpendicular to $\pi$.

Points A(2, –1, 2), B(4, –4, 2), C(5, –6, 3) lie on $\pi$.

**(ii)** Write down the vectors $\overrightarrow{AB}$ and $\overrightarrow{AC}$.

**(iii)** Use scalar products to obtain two equations for $a$ and $b$.

**(iv)** Find the equation of the plane $\pi$.

**(v)** Find the angle which the plane $\pi$ makes with the plane $x = 0$.

**(vi)** Point D is the mid-point of AC. Point E is on the line between D and B such that DE : EB = 1 : 2. Find the co-ordinates of E.

[MEI]

**22**

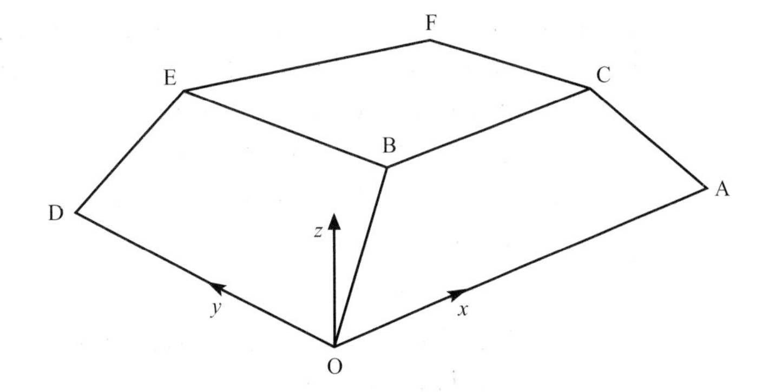


The diagram, which is not to scale, illustrates part of the roof of a building. Lines OA and OD are horizontal and at right angles. Lines BC and BE are also horizontal and at right angles. Line BC is parallel to OA and BE is parallel to OD.

Axes are taken with O as origin, the $x$ axis along OA, the $y$ axis along OD and the $z$ axis vertically upwards. The units are metres.

Point A has the co-ordinates (50, 0, 0) and point D has the co-ordinates (0, 20, 0). The equation of line OB is $\frac{x}{4} = \frac{y}{3} = \frac{z}{2}$. The equation of plane CBEF is $z = 3$.

**(i)** Find the co-ordinates of B.

**(ii)** Verify that the equation of plane AOBC is $2y - 3z = 0$.

**(iii)** Find the equation of plane DOBE.

**(iv)** Write down normal vectors for planes AOBC and DOBE. Find the angle between these normal vectors. Hence write down the internal angle between the two roof surfaces AOBC and DOBE.

[MEI]

**23** ABCD is a parallelogram. The co-ordinates of A, B, D are (4, 2, 3), (18, 4, 8) and (–1, 12, 13) respectively. The origin of co-ordinates is O.

**(i)** Find the vectors $\overrightarrow{AB}$ and $\overrightarrow{AD}$. Find the co-ordinates of C.

**(ii)** Show that ABCD is a square of side 15 units.

**(iii)** Show that $\overrightarrow{OA}$ can be expressed in the form $\lambda\overrightarrow{AB} + \mu\overrightarrow{AD}$, stating the values of $\lambda$ and $\mu$. What does this tell you about the plane ABCD?

**(iv)** Find the cartesian equation of the plane ABCD.

[MEI]

**24** A tunnel is to be excavated through a hill. In order to define position, co-ordinates $(x, y, z)$ are taken relative to an origin O such that $x$ is the distance east from O, $y$ is the distance north and $z$ is the vertical distance upwards, with one unit equal to 100 m.

The tunnel starts at point A(2, 3, 5) and runs in the direction $\begin{pmatrix} 1 \\ 1 \\ -0.5 \end{pmatrix}$.

It meets the hillside again at B. At B the side of the hill forms a plane with equation $x + 5y + 2z = 77$.

**(i)** Write down the equation of the line AB in the form $\mathbf{r} = \mathbf{u} + \lambda\mathbf{t}$.

**(ii)** Find the co-ordinates of B.

**(iii)** Find the angle which AB makes with the upward vertical.

**(iv)** An old tunnel through the hill has equation $\mathbf{r} = \begin{pmatrix} 4 \\ 1 \\ 2 \end{pmatrix} + \mu\begin{pmatrix} 7 \\ 15 \\ 0 \end{pmatrix}$.

Show that the point P on AB where $x = 7\frac{1}{2}$ is directly above a point Q in the old tunnel.

Find the vertical separation PQ of the tunnels at this point.

[MEI]

## INVESTIGATIONS

### MAGIC EYE

You may well have seen other pictures like that in figure 5.36. Although it is nothing more than a collection of marks on a flat sheet of paper, your eyes can be tricked into seeing it as a three-dimensional object at some distance beyond the page. As shown in figure 5.35, the principle is very simple.

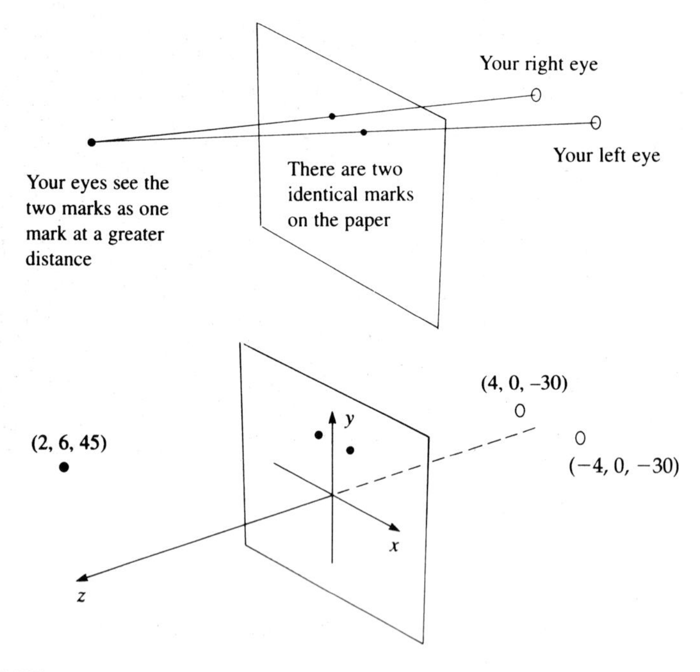


**Figure 5.35**

Take the paper as the $xy$ plane, that is the plane $z = 0$, with the origin at the centre of the paper and 1 cm to represent 1 unit.

Taking the positions of your eyes to be (4, 0, –30) and (–4, 0, –30), find the positions on the paper of the two points needed to produce a single image at the point (2, 6, 45).

Design a simple 'magic eye' of your own.

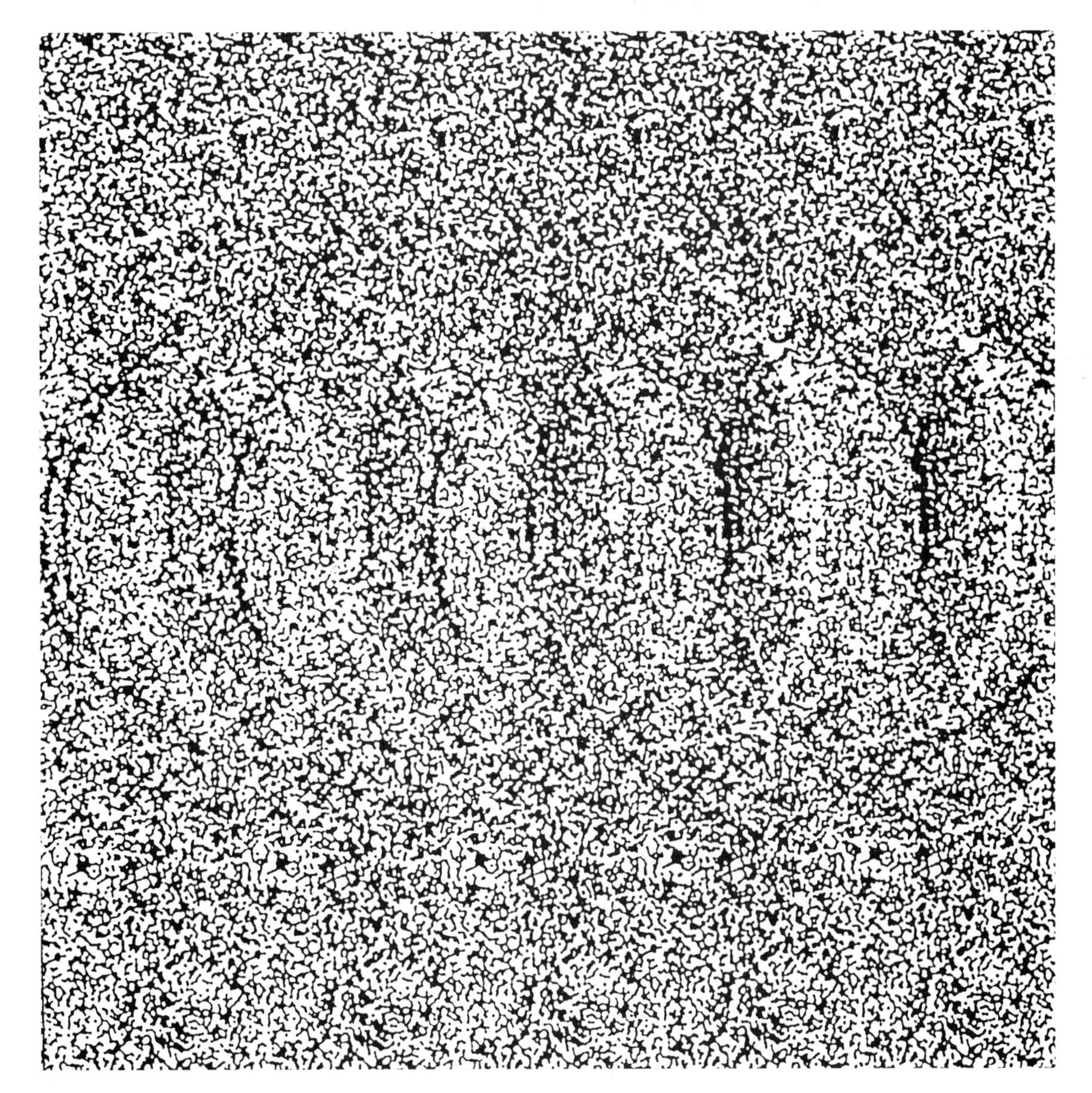

**Figure 5.36**



## KEY POINTS

1. A vector quantity has magnitude and direction.
2. A scalar quantity has magnitude only.
3. Vectors are typeset in bold, **a** or **OA**, or in the form $\overrightarrow{OA}$. They are handwritten either in the underlined form $\underset{\sim}{a}$, or as $\overrightarrow{OA}$.
4. The length (or modulus or magnitude) of the vector **a** is written as $a$ or as $|\mathbf{a}|$.

**5** Unit vectors in the $x$, $y$ and $z$ directions are denoted by **i**, **j** and **k**, respectively.

**6** A vector may be specified in

- Magnitude–direction form: $(r, \theta)$ (in two dimensions)
- Component form: $x\mathbf{i} + y\mathbf{j}$ or $\begin{pmatrix} x \\ y \end{pmatrix}$ (in two dimensions)

$x\mathbf{i} + y\mathbf{j} + z\mathbf{k}$ or $\begin{pmatrix} x \\ y \\ z \end{pmatrix}$ (in three dimensions)

**7** The position vector $\overrightarrow{OP}$ of a point P is the vector joining the origin to P.

**8** The vector $\overrightarrow{AB}$ is $\mathbf{b} - \mathbf{a}$, where **a** and **b** are the position vectors of A and B.

**9** The vector **r** often denotes the position vector of a general point.

**10** The vector equation of the line through A with direction vector **u** is given by

$$\mathbf{r} = \mathbf{a} + \lambda\mathbf{u}$$

**11** The vector equation of the line through points A and B is given by

$$\begin{aligned} \mathbf{r} &= \overrightarrow{OA} + \lambda\overrightarrow{AB} \\ &= \mathbf{a} + \lambda(\mathbf{b} - \mathbf{a}) \\ &= (1 - \lambda)\mathbf{a} + \lambda\mathbf{b} \end{aligned}$$

**12** The equation of the line through $(a_1, a_2, a_3)$ in the direction $\begin{pmatrix} u_1 \\ u_2 \\ u_3 \end{pmatrix}$ is given by

$$\mathbf{r} = \begin{pmatrix} a_1 \\ a_2 \\ a_3 \end{pmatrix} + \lambda\begin{pmatrix} u_1 \\ u_2 \\ u_3 \end{pmatrix} \quad \text{vector form}$$

$$\frac{x - a_1}{u_1} = \frac{y - a_2}{u_2} = \frac{z - a_3}{u_3} \quad \text{cartesian form.}$$

**13** The angle between two vectors, **a** and **b**, is given by $\theta$ in

$$\cos\theta = \frac{\mathbf{a}.\mathbf{b}}{|\mathbf{a}|\,|\mathbf{b}|}$$

where $\mathbf{a}.\mathbf{b} = a_1b_1 + a_2b_2$ (in two dimensions)
$= a_1b_1 + a_2b_2 + a_3b_3$ (in three dimensions).

**14** The vector equation of the plane through the points A, B and C is

$$\mathbf{r} = \overrightarrow{OA} + \lambda\overrightarrow{AB} + \mu\overrightarrow{AC}.$$

**15** The cartesian equation of a plane perpendicular to the vector $\mathbf{n} = \begin{pmatrix} n_1 \\ n_2 \\ n_3 \end{pmatrix}$ is $n_1x + n_2y + n_3z + d = 0$.

**16** The equation of the plane through the point with position vector **a**, and perpendicular to **n**, is given by $(\mathbf{r} - \mathbf{a}).\,\mathbf{n} = 0$.

# 6 Differential equations

**The greater our knowledge increases, the more our ignorance unfolds.**

*John F. Kennedy*

Suppose you are in a hurry to go out and want to drink a cup of hot tea before you go.

How long will you have to wait until it is cool enough to drink?

To solve this problem, you would need to know something about the rate at which liquids cool at different temperatures. Figure 6.1 shows an example of the temperature of a liquid plotted against time.

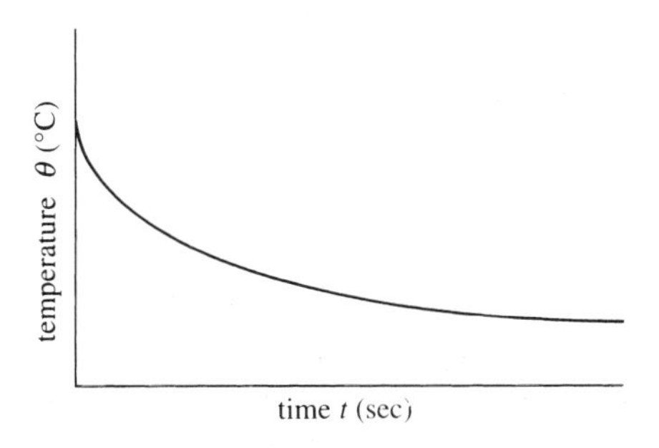


**Figure 6.1**

Notice that the graph is steepest at high temperatures and becomes less steep as the liquid cools. In other words, the rate of change of temperature is numerically greatest at high temperatures and gets numerically less as the temperature drops. The rate of change is always negative since the temperature is decreasing.

If you study physics, you may have come across Newton's law of cooling: The rate of cooling of a body is proportional to the temperature of the body above that of the surrounding air.

The gradient of the temperature graph may be written as $\frac{\mathrm{d}\theta}{\mathrm{d}t}$, where $\theta$ is the temperature of the liquid, and $t$ is the time. The quantity $\frac{\mathrm{d}\theta}{\mathrm{d}t}$ tells us the rate at which the temperature of the liquid is increasing. As the liquid is cooling, $\frac{\mathrm{d}\theta}{\mathrm{d}t}$ will be negative, so the rate of cooling may be written as $-\frac{\mathrm{d}\theta}{\mathrm{d}t}$.

The temperature of the liquid above that of the surrounding air may be written as $\theta - \theta_0$, where $\theta_0$ is the temperature of the surrounding air. So Newton's law of cooling may be expressed mathematically as:

$$-\frac{d\theta}{dt} \propto (\theta - \theta_0)$$

or $$\frac{d\theta}{dt} = -k(\theta - \theta_0)$$

where $k$ is a positive constant.

Any equation, like this one, which involves a derivative, such as $\frac{d\theta}{dt}$, $\frac{dy}{dx}$ or $\frac{d^2y}{dx^2}$, is known as a *differential equation.* A differential equation which only involves a first derivative such as $\frac{dy}{dx}$ is called a *first-order differential equation.* One which involves a second derivative such as $\frac{d^2y}{dx^2}$ is called a *second-order differential equation.* A third-order differential equation involves a third derivative and so on. In this chapter, you will be looking only at first-order differential equations such as the one above for Newton's law of cooling.

By the end of this chapter, you will be able to solve problems such as the tea cooling problem given at the beginning of this chapter, by using first-order differential equations.

## Forming differential equations from rates of change

If you are given sufficient information about the rate of change of a quantity, such as temperature or velocity, you can work out a differential equation to model the situation, like the one above for Newton's law of cooling. It is important to look carefully at the wording of the problem which you are studying in order to write an equivalent mathematical statement. For example, if the altitude of an aircraft is being considered, the phrase 'the rate of change of height' might be used. This actually means 'the rate of change of height *with respect to time*' and could be written as $\frac{dh}{dt}$. However, we might be more interested in how the height of the aircraft changes according to the horizontal distance it has travelled. In this case, we would talk about 'the rate of change of height *with respect to horizontal distance*' and could write this as $\frac{dh}{dx}$, where $x$ is the horizontal distance travelled.

Some of the situations you meet in this chapter involve motion along a straight line, and so you will need to know the meanings of the associated terms.

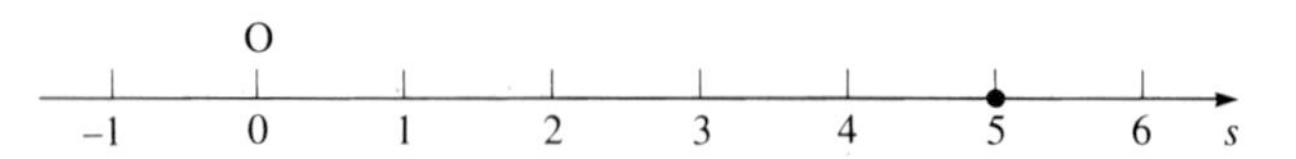


**Figure 6.2**

The position of an object (+5 in figure 6.2) is its distance from the origin O in the direction you have chosen to define as being positive.

The rate of change of position of the object with respect to time is its velocity, and this can take positive or negative values according to whether the object is moving away from the origin or towards it.

$$v = \frac{\mathrm{d}s}{\mathrm{d}t}$$

The rate of change of an object's velocity with respect to time is called its acceleration, $a$.

$$a = \frac{\mathrm{d}v}{\mathrm{d}t}$$

Velocity and acceleration are vector quantities but in one-dimensional motion there is no choice in direction, only in sense (i.e. whether positive or negative). Consequently, as you may already have noticed, we have chosen not to use the conventional bold type for vectors in this chapter.

**EXAMPLE 6.1**

An object is moving through a liquid so that the rate at which its velocity decreases is proportional to its velocity at any given instant. When it enters the liquid, it has a velocity of 5 $\mathrm{ms}^{-1}$ and the velocity is decreasing at a rate of 1 $\mathrm{ms}^{-2}$. Find the differential equation to model this situation.

**SOLUTION**

The rate of change of velocity means the rate of change of velocity with respect to time and so can be written as $\frac{\mathrm{d}v}{\mathrm{d}t}$. As it is decreasing, the rate of change must be negative, so

$$-\frac{\mathrm{d}v}{\mathrm{d}t} \propto v$$

or $$\frac{\mathrm{d}v}{\mathrm{d}t} = -kv$$

where $k$ is a positive constant.

When the object enters the liquid its velocity is 5 $\mathrm{ms}^{-1}$, so $v = 5$, and the velocity is decreasing at the rate of 1 $\mathrm{ms}^{-2}$, so

$$\frac{\mathrm{d}v}{\mathrm{d}t} = -1.$$

Putting this information into the equation gives

$$-1 = -k \times 5 \quad \Rightarrow \quad k = \tfrac{1}{5}.$$

So the situation is modelled by the differential equation

$$\frac{\mathrm{d}v}{\mathrm{d}t} = -\frac{v}{5}.$$

**EXAMPLE 6.2**

A model is proposed for the temperature gradient within a star, in which the temperature decreases with respect to the distance from the centre of the star at a rate which is inversely proportional to the square of the distance from the centre. Express this model as a differential equation.

**SOLUTION**

In this example the rate of change of temperature is not with respect to time but with respect to distance. If $\theta$ represents the temperature of the star and $r$ the distance from the centre of the star, the rate of change of temperature with respect to distance may be written as $-\frac{d\theta}{dr}$, so

$$-\frac{d\theta}{dr} \propto \frac{1}{r^2} \quad \text{or} \quad \frac{d\theta}{dr} = -\frac{k}{r^2}$$

where $k$ is a positive constant.

*Note*

This model must break down near the centre of the star, otherwise it would be infinitely hot there.

**EXAMPLE 6.3**

The area $A$ of a square is increasing at a rate proportional to the length of its side $s$. Find an expression for $\frac{ds}{dt}$.

**SOLUTION**

The rate of increase of $A$ with respect to time may be written as $\frac{dA}{dt}$. As this is proportional to $s$, we can write

$$\frac{dA}{dt} = ks$$

where $k$ is a positive constant.

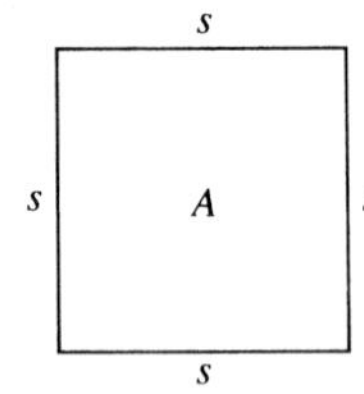

**Figure 6.3**

We can use the chain rule to write down an expression for $\frac{ds}{dt}$ in terms of $\frac{dA}{dt}$:

$$\frac{ds}{dt} = \frac{ds}{dA} \times \frac{dA}{dt}.$$

We now need an expression for $\frac{ds}{dA}$. Because A is a square

$$A = s^2$$

$$\Rightarrow \quad \frac{dA}{ds} = 2s$$

$$\Rightarrow \quad \frac{ds}{dA} = \frac{1}{2s}.$$

Substituting the expressions for $\frac{ds}{dA}$ and $\frac{dA}{dt}$ into the expression for $\frac{ds}{dt}$:

$$\Rightarrow \quad \frac{ds}{dt} = \frac{1}{2s} \times ks$$

$$\Rightarrow \quad \frac{ds}{dt} = \tfrac{1}{2}k.$$

## EXERCISE 6A

**1** The differential equation

$$\frac{dv}{dt} = 5v^2$$

models the motion of a particle, where $v$ is the velocity of the particle in $\text{ms}^{-1}$ and $t$ is the time in seconds. Explain the meaning of $\frac{dv}{dt}$ and what the differential equation tells us about the motion of the particle.

**2** A spark from a Roman candle is moving in a straight line at a speed which is inversely proportional to the square of the distance which the spark has travelled from the candle. Find an expression for the speed (i.e. the rate of change of distance travelled) of the spark.

**3** The rate at which a sunflower increases in height is proportional to the natural logarithm of the difference between its final height $H$ and its height $h$ at a particular time. Find a differential equation to model this situation.

**4** In a chemical reaction in which substance A is converted into substance B, the rate of increase of the mass of substance B is inversely proportional to the mass of substance B present. Find a differential equation to model this situation.

**5** After a major advertising campaign, an engineering company finds that its profits are increasing at a rate proportional to the square root of the profits at any given time. Find an expression to model this situation.

**6** The coefficient of restitution $e$ of a squash ball increases with respect to the ball's temperature $\theta$ at a rate proportional to the temperature, for typical playing temperatures. (The coefficient of restitution is a measure of how elastic, or bouncy, the ball is. Its value lies between 0 and 1, 0 meaning that the ball is not at all elastic and 1 meaning that it is perfectly elastic.) Find a differential equation to model this situation.

**7** A cup of tea cools at a rate proportional to the temperature of the tea above that of the surrounding air. Initially, the tea is at a temperature of 95°C and is cooling at a rate of $0.5°C\ s^{-1}$. The surrounding air is at 15°C. Find a differential equation to model this situation.

**8** The rate of increase of bacteria is modelled as being proportional to the number of bacteria at any time during their initial growth phase.

When the bacteria number $2 \times 10^6$ they are increasing at a rate of $10^5$ per day. Find a differential equation to model this situation.

**9** The acceleration (i.e. the rate of change of velocity) of a moving object under a particular force is inversely proportional to the square root of its velocity. When the speed is $4\ ms^{-1}$ the acceleration is $2\ ms^{-2}$. Find a differential equation to model this situation.

**10** The radius of a circular ink blot is increasing at a rate inversely proportional to its area $A$. Find an expression for $\frac{dA}{dt}$.

**11** A poker, 80 cm long, has one end in a fire. The temperature of the poker decreases with respect to the distance from that end at a rate proportional to that distance. Halfway along the poker, the temperature is decreasing at a rate of $10°C\ cm^{-1}$. Find a differential equation to model this situation.

**12** A conical egg timer, shown in the diagram, is letting sand through from top to bottom at a rate of $0.02\ cm^3\ s^{-1}$. Find an expression for the rate of change of height $\left(\frac{dh}{dt}\right)$ of the sand in the top of the timer.

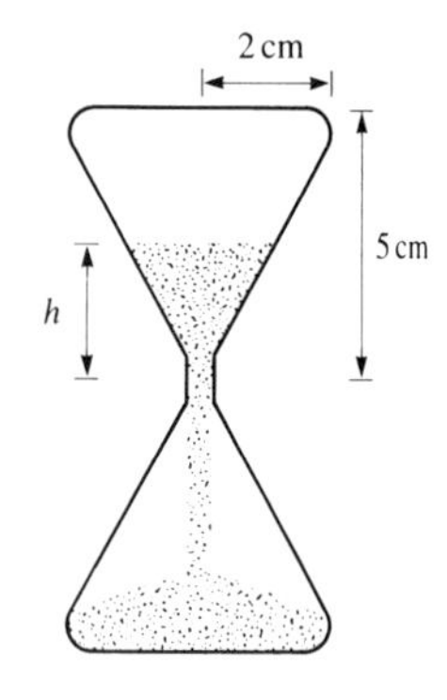


**13** A spherical balloon is allowed to deflate. The rate at which air is leaving the balloon is proportional to the volume $V$ of air left in the balloon. When the radius of the balloon is 15 cm, air is leaving at a rate of $8\ cm^3\ s^{-1}$. Find an expression for $\frac{dV}{dt}$.

**14** A tank is shaped as a cuboid with a square base of side 10 cm. Water runs out through a hole in the base at a rate proportional to the square root of the height, $h$ cm, of water in the tank. At the same time, water is pumped into the tank at a constant rate of $2\ cm^3\ s^{-1}$. Find an expression for $\frac{dh}{dt}$.

Figure 6.4 shows the isobars (lines of equal pressure) on a weather map featuring a storm. The wind direction is almost parallel to the isobars and its speed is proportional to the pressure gradient.

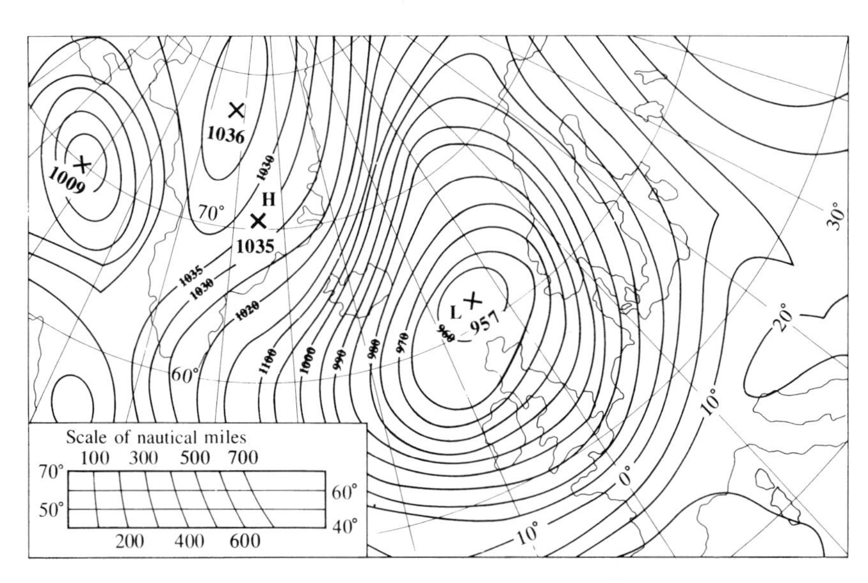


**Figure 6.4**

Draw a line from the point H to the point L. This runs approximately perpendicular to the isobars. It is suggested that along this line the pressure gradient (and so the wind speed) may be modelled by the differential equation

$$\frac{dp}{dx} = -a\sin bx$$

Suggest values for $a$ and $b$, and comment on the suitability of this model.

## Solving differential equations

Finding an expression for $f(x)$ from a differential equation involving derivatives of $f(x)$ is called solving the equation.

Some differential equations may be solved simply by integration.

**EXAMPLE 6.4**

Solve the differential equation

$$\frac{dy}{dx} = 3x^2 - 2.$$

**SOLUTION**

Integrating gives

$$y = \int (3x^2 - 2)\, dx$$

$$y = x^3 - 2x + c.$$

**THE GENERAL SOLUTION OF THE DIFFERENTIAL EQUATION**

Notice that when you solve a differential equation, you get not just one solution, but a whole family of solutions, as $c$ can take any value. This is called the *general solution* of the differential equation. The family of solutions for the differential equation in the example above would be translations in the $y$ direction of the curve $y = x^3 - 2x$. Graphs of members of the family of curves can be found on page 189.

## The method of separation of variables

It is not difficult to solve a differential equation like the one in Example 6.4, because the right-hand side is a function of $x$ only. So long as the function can be integrated, the equation can be solved.

Now consider the differential equation

$$\frac{dy}{dx} = xy.$$

This cannot be solved directly by integration, because the right-hand side is a function of both $x$ and $y$. However, as you will see in the next example, you can solve this and similar differential equations where the right-hand side consists of a function of $x$ and a function of $y$ multiplied together.

**EXAMPLE 6.5**

Find, for $y > 0$, the general solution of the differential equation

$$\frac{dy}{dx} = xy.$$

**SOLUTION**

The equation may be rewritten as

$$\frac{1}{y}\frac{dy}{dx} = x$$

so that the right-hand side is now a function of $x$ only.

Integrating both sides with respect to $x$ gives

$$\int \frac{1}{y}\frac{dy}{dx}\,dx = \int x\,dx$$

As $\dfrac{dy}{dx}\,dx$ can be written as $dy$

$$\int \frac{1}{y}\,dy = \int x\,dx$$

Both sides may now be integrated separately:

$$\ln|y| = \tfrac{1}{2}x^2 + c.$$

Since we have been told $y > 0$, we may drop the modulus symbol. In this case, $|y| = y$

(There is no need to put a constant of integration on both sides of the equation.)

We now need to rearrange the solution above to give $y$ in terms of $x$. Making both sides power of e gives

$$e^{\ln y} = e^{\frac{1}{2}x^2 + c}$$

$$\Rightarrow \quad y = e^{\frac{1}{2}x^2 + c}$$

Notice that the right-hand side is $e^{\frac{1}{2}x^2 + c}$ and not $e^{\frac{1}{2}x^2} + e^c$

which is rearranged to give

$$y = e^{\frac{1}{2}x^2}e^c.$$

This expression can be simplified by replacing $e^c$ with a new constant $A$.

So

$$y = Ae^{\frac{1}{2}x^2}.$$

***Note***

Usually the first part of this process is carried out in just one step

$$\frac{dy}{dx} = xy$$

and can immediately be rewritten as

$$\int \frac{1}{y}\,dy = \int x\,dx.$$

This method is called *separation of variables.* It can be helpful to do this by thinking of the differential equation as though $\dfrac{dy}{dx}$ were a fraction, and trying to rearrange the equation to obtain all the $x$ terms on one side and all the $y$ terms on the other. Then just insert an integration sign on each side. Remember that $dy$ and $dx$ must both end up on the top line (numerator).

**EXAMPLE 6.6**

Find the general solution of the differential equation

$$\frac{dy}{dx} = e^{-y}.$$

**SOLUTION**

Separating the variables (see the warning below) gives

$$\int \frac{1}{e^{-y}}\,dy = \int dx$$

which is rearranged to give

$$\int e^{y}\,dy = \int dx.$$

The right-hand side can be thought of as integrating 1 with respect to $x$:

$$e^{y} = x + c.$$

Taking logarithms of both sides:

$$y = \ln(x + c).$$

$\ln(x + c)$ is not the same as $\ln x + c$.

**EXERCISE 6B**

**1** Solve the following differential equations by integration.

**(i)** $\frac{dy}{dx} = x^2$ **(ii)** $\frac{dy}{dx} = \cos x$

**(iii)** $\frac{dy}{dx} = e^{x}$ **(iv)** $\frac{dy}{dx} = \sqrt{x}$

**2** Find the general solutions of the following differential equations by separating the variables.

**(i)** $\frac{dy}{dx} = xy^2$ **(ii)** $\frac{dy}{dx} = \frac{x^2}{y}$ **(iii)** $\frac{dy}{dx} = y$

**(iv)** $\frac{dy}{dx} = e^{x-y}$ **(v)** $\frac{dy}{dx} = \frac{y}{x}$ **(vi)** $\frac{dy}{dx} = x\sqrt{y}$

**(vii)** $\frac{dy}{dx} = y^2 \cos x$ **(viii)** $\frac{dy}{dx} = \frac{x(y^2+1)}{y(x^2+1)}$ **(ix)** $\frac{dy}{dx} = xe^{y}$

**(x)** $\frac{dy}{dx} = \frac{x\ln x}{y^2}$

## Particular solutions

You have already seen that a differential equation has an infinite number of different solutions corresponding to different values of the constant of integration. In Example 6.4, we found that

$$\frac{dy}{dx} = 3x^2 - 2$$

had a general solution of $y = x^3 - 2x + c$.

Figure 6.5 shows the curves of the solutions corresponding to some different values of $c$.

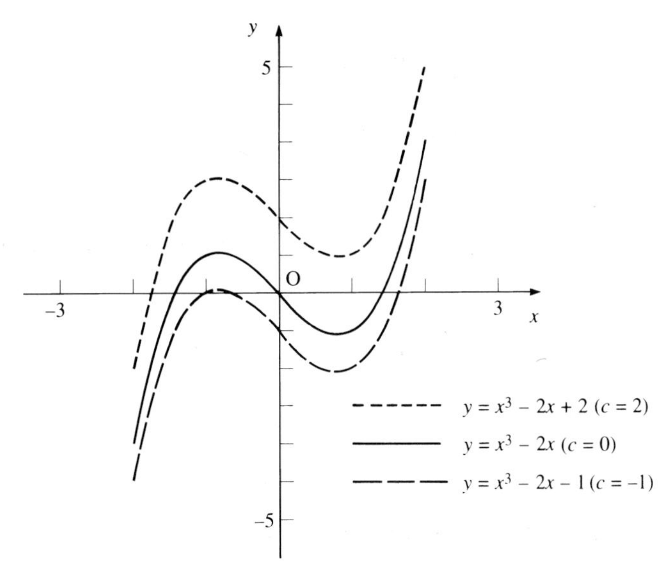


**Figure 6.5**

If you are given some more information, you can find out which of the possible solutions is the one that matches the situation in question. For example, you might be told that when $x = 1$, $y = 0$. This tells you that the correct solution is the one whose curve passes through the point (1, 0). You can use this information to find out the value of $c$ for this particular solution by substituting the values $x = 1$ and $y = 0$ into the general solution:

$$y = x^3 - 2x + c$$

$$0 = 1 - 2 + c$$

$$\Rightarrow \quad c = 1.$$

So the solution in this case is $y = x^3 - 2x + 1$.

This is called the *particular solution.*

**EXAMPLE 6.7**

**(i)** Find the general solution of the differential equation $\dfrac{dy}{dx} = y^2$.

**(ii)** Find the particular solution for which $y = 1$ when $x = 0$.

**SOLUTION**

**(i)** Separating the variables gives $\int \frac{1}{y^2}\, dy = \int dx$

$$-\frac{1}{y} = x + c.$$

The general solution is $\quad y = -\dfrac{1}{x + c}$.

Figure 6.6 shows the set of solution curves.

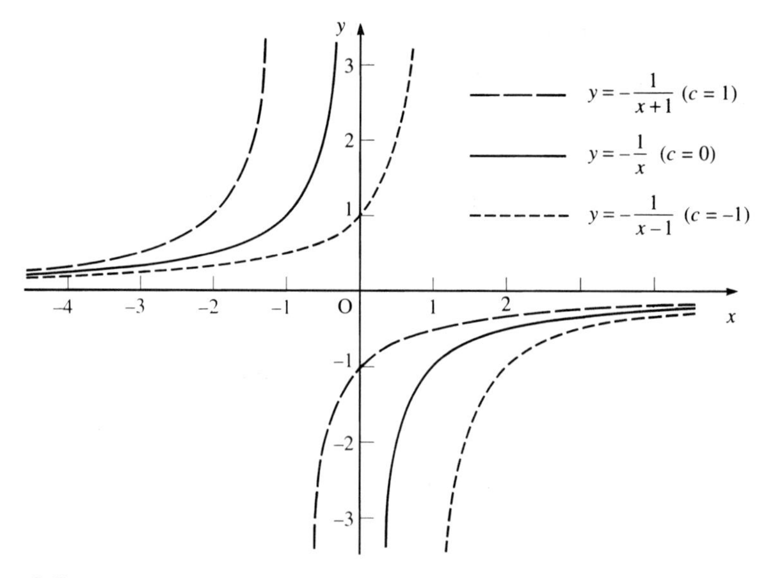


**Figure 6.6**

**(ii)** When $x = 0$, $y = 1$, which gives

$$1 = -\frac{1}{c} \quad \Rightarrow \quad c = -1.$$

So the particular solution is

$$y = -\frac{1}{x - 1} \quad \text{or} \quad y = \frac{1}{1 - x}.$$

This is one of the curves illustrated in figure 6.6.

**EXAMPLE 6.8**

The acceleration of an object is inversely proportional to its velocity at any given time and the direction of motion is taken to be positive. When the velocity is 1 $\text{ms}^{-1}$, the acceleration is 3 $\text{ms}^{-2}$.

**(i)** Find a differential equation to model this situation.

**(ii)** Find the particular solution to this differential equation for which the initial velocity is 2 $\text{ms}^{-1}$.

**(iii)** In this case, how long does the object take to reach a velocity of 8 $\text{ms}^{-1}$?

**SOLUTION**

**(i)** $\dfrac{\mathrm{d}v}{\mathrm{d}t} = \dfrac{k}{v}$

When $v = 1$, $\dfrac{\mathrm{d}v}{\mathrm{d}t} = 3$ so $k = 3$, which gives $\dfrac{\mathrm{d}v}{\mathrm{d}t} = \dfrac{3}{v}$.

**(ii)** Separating the variables:

$$\int v\,\mathrm{d}v = \int 3\,\mathrm{d}t$$

$$\tfrac{1}{2}v^2 = 3t + c.$$

When $t = 0$, $v = 2$ so $c = 2$, which gives

$$\tfrac{1}{2}v^2 = 3t + 2$$

$$v^2 = 6t + 4.$$

Since the direction of motion is positive

$$v = \sqrt{6t + 4}.$$

**(iii)** When $v = 8$ $\quad 64 = 6t + 4$

$$60 = 6t \quad \Rightarrow \quad t = 10.$$

The object takes 10 seconds.

The graph of the particular solution is shown in figure 6.7.

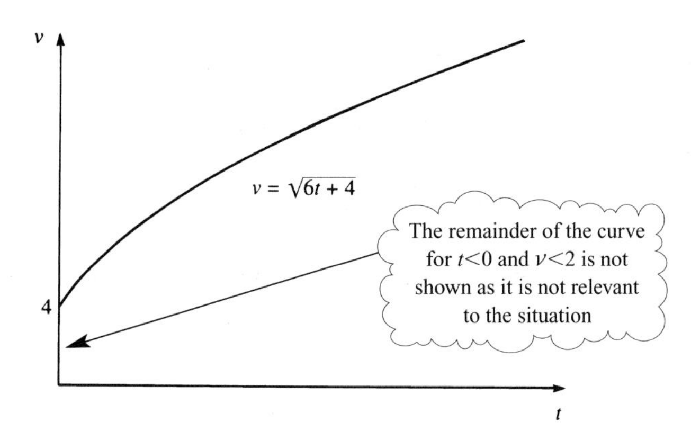


**Figure 6.7**

**EXERCISE 6C**

**1** Find the particular solution of each of the following differential equations.

**(i)** $\dfrac{dy}{dx} = x^2 - 1$ $\quad y = 2$ when $x = 3$

**(ii)** $\dfrac{dy}{dx} = x^2y$ $\quad y = 1$ when $x = 0$

**(iii)** $\dfrac{dy}{dx} = xe^{-y}$ $\quad y = 0$ when $x = 0$

**(iv)** $\dfrac{dy}{dx} = y^2$ $\quad y = 1$ when $x = 1$

**(v)** $\dfrac{dy}{dx} = x(y + 1)$ $\quad y = 0$ when $x = 1$

**(vi)** $\dfrac{dy}{dx} = y^2 \sin x$ $\quad y = 1$ when $x = 0$

**2** A cold liquid at temperature $\theta$ °C, where $\theta < 20$, is standing in a warm room. The temperature of the liquid obeys the differential equation

$$\frac{d\theta}{dt} = 2(20 - \theta)$$

where the time $t$ is measured in seconds.

**(i)** Find the general solution of this differential equation.

**(ii)** Find the particular solution for which $\theta = 5$ when $t = 0$.

**(iii)** In this case, how long does the liquid take to reach a temperature of 18°C?

**3** A population of rabbits increases so that the number of rabbits $N$ (in hundreds), after $t$ years is modelled by the differential equation

$$\frac{dN}{dt} = N.$$

**(i)** Find the general solution for $N$ in terms of $t$.

**(ii)** Find the particular solution for which $N = 10$ when $t = 0$.

**(iii)** What will happen to the number of rabbits when $t$ becomes very large? Why is this not a realistic model for an actual population of rabbits?

**4** An object is moving so that its velocity $v\left(= \dfrac{ds}{dt}\right)$ is inversely proportional to its displacement $s$ from a fixed point. If its velocity is 1 $\text{ms}^{-1}$ when its displacement is 2 m, find a differential equation to model the situation. Find the general solution of your differential equation.

**5 (i)** Write $\dfrac{1}{y(3-y)}$ in partial fractions.

**(ii)** Find $\displaystyle\int \frac{1}{y(3-y)}\,\mathrm{d}y$.

**(iii)** Solve the differential equation

$$x\frac{\mathrm{d}y}{\mathrm{d}x} = y(3-y)$$

where $x = 2$ when $y = 2$, giving $y$ as a function of $x$.

[MEI]

**6** Given that $k$ is a constant, find the solution of the differential equation

$$\frac{\mathrm{d}y}{\mathrm{d}t} + ky = 2k$$

for which $y = 3$ when $t = 0$.

Sketch the graph of $y$ against $|kt|$, making clear how it behaves for large values of $|kt|$.

[MEI]

**7** A colony of bacteria which is initially of size 1500 increases at a rate proportional to its size so that, after $t$ hours, its population $N$ satisfies the equation

$$\frac{\mathrm{d}N}{\mathrm{d}t} = kN.$$

**(i)** If the size of the colony increases to 3000 in 20 hours, solve the differential equation to find $N$ in terms of $t$.

**(ii)** What size is the colony when $t = 80$?

**(iii)** How long did it take, to the nearest minute, for the population to increase from 2000 to 3000?

[MEI]

**8 (i)** Show that

$$\frac{x^2+1}{x^2-1} = 1 + \frac{2}{x^2-1}.$$

**(ii)** Find the partial fractions for

$$\frac{2}{(x-1)(x+1)}.$$

**(iii)** Solve the differential equation

$$(x^2-1)\frac{\mathrm{d}y}{\mathrm{d}x} = -(x^2+1)y \qquad \text{(where } x > 1\text{)}$$

given that $y = 1$ when $x = 3$. Express $y$ as a function of $x$.

[MEI]

**9** A hemispherical bowl of radius $a$ has its axis vertical and is full of water. At time $t = 0$ water starts running out of a small hole in the bottom of the bowl so that the depth of water in the bowl at time $t$ is $x$. The rate at which the volume of water is decreasing is proportional to $x$. Given that the volume of water in the bowl when the depth is $x$ is $\pi(ax^2 - \frac{1}{3}x^3)$, show that there is a positive constant $k$ such that

$$\pi(2ax - x^2)\frac{dx}{dt} = -kx.$$

Given that the bowl is empty after a time $T$, show that

$$k = \frac{3\pi a^2}{2T}.$$

[MEI]

**10** The square horizontal cross-section of a container has side 2 m. Water is poured in at the constant rate of $0.08\ \text{m}^3\text{s}^{-1}$ and, at the same time, leaks out of a hole in the base at the rate of $0.12x\ \text{m}^3\text{s}^{-1}$, where $x$ m is the depth of the water in the container at time $t$ s. So the volume, $V\ \text{m}^3$, of the water in the container at time $t$ is given by $V = 4x$ and the rate of change of volume is given by

$$\frac{dV}{dt} = 0.08 - 0.12x.$$

Use these results to find an equation for $\frac{dx}{dt}$ in terms of $x$ and solve this to find $x$ in terms of $t$ if the container is empty initially.

Determine to the nearest 0.1 s the time taken for the depth to rise from 0.1 to 0.5 m.

[MEI]

**11** A patch of oil pollution in the sea is approximately circular in shape. When first seen its radius was 100 m and its radius was increasing at a rate of 0.5 m per minute. At a time $t$ minutes later, its radius is $r$ metres. An expert believes that, if the patch is untreated, its radius will increase at a rate which is proportional to $\frac{1}{r^2}$.

**(i)** Write down a differential equation for this situation, using a constant of proportionality, $k$.

**(ii)** Using the initial conditions, find the value of $k$. Hence calculate the expert's prediction of the radius of the oil patch after 2 hours.

The expert thinks that if the oil patch is treated with chemicals then its radius will increase at a rate which is proportional to $\frac{1}{r^2(2 + t)}$.

**(iii)** Write down a differential equation for this new situation and, using the same initial conditions as before, find the value of the new constant of proportionality.

**(iv)** Calculate the expert's prediction of the radius of the treated oil patch after 2 hours.

[MEI]

**12** To control the pests inside a large greenhouse, 600 ladybirds were introduced. After $t$ days there are $P$ ladybirds in the greenhouse.

In a simple model, $P$ is assumed to be a continuous variable satisfying the differential equation

$$\frac{dP}{dt} = kP, \text{ where } k \text{ is a constant.}$$

**(i)** Solve the differential equation, with initial condition $P = 600$ when $t = 0$, to express $P$ in terms of $k$ and $t$.

Observations of the number of ladybirds (estimated to the nearest hundred) were made as follows:

| $t$ | 0 | 150 | 250 |
|---|---|---|---|
| $P$ | 600 | 1200 | 3100 |

**(ii)** Show that $P = 1200$ when $t = 150$ implies that $k \approx 0.004\,62$. Show that this is not consistent with the observed value when $t = 250$.

In a refined model, allowing for seasonal variations, it is assumed that $P$ satisfies the differential equation

$$\frac{dP}{dt} = P\,[0.005 - 0.008\cos(0.02t)]$$

with initial condition $P = 600$ when $t = 0$.

**(iii)** Solve this differential equation to express $P$ in terms of $t$, and comment on how well this fits with the data given above.

**(iv)** Show that, according to the refined model, the number of ladybirds will decrease initially, and find the smallest number of ladybirds in the greenhouse.

[MEI]

**13 (i)** Express $\dfrac{1}{(2 - x)(1 + x)}$ in partial fractions.

An industrial process creates a chemical C. At time $t$ hours after the start of the process the amount of C produced is $x$ kg. The rate at which C is produced is given by the differential equation

$$\frac{dx}{dt} = k(2 - x)(1 + x)e^{-t},$$

where $k$ is a constant.

**(ii)** When $t = 0$, $x = 0$ and the rate of production of C is $\frac{2}{3}$ kg per hour. Calculate the value of $k$.

**(iii)** Show that

$$\ln\left(\frac{1+x}{2-x}\right) = -e^{-t} + 1 - \ln 2,$$

provided that $x < 2$.

**(iv)** Find, in hours, the time taken to produce 0.5 kg of C, giving your answer correct to 2 decimal places.

**(v)** Show that there is a finite limit to the amount of C which this process can produce, however long it runs, and determine the value of this limit.

[MEI]

**14 (i)** Use integration by parts to evaluate

$$\int 4x\cos 2x \, dx.$$

**(ii)** Use part (i), together with a suitable expression for $\cos^2 x$, to show that

$$\int 8x\cos^2 x \, dx = 2x^2 + 2x\sin 2x + \cos 2x + c.$$

**(iii)** Find the solution of the differential equation

$$\frac{dy}{dx} = \frac{8x\cos^2 x}{y}$$

which satisfies $y = \sqrt{3}$ when $x = 0$.

**(iv)** Show that any point $(x, y)$ on the graph of this solution which satisfies $\sin 2x = 1$ also lies on one of the lines $y = 2x + 1$ or $y = -2x - 1$.

[MEI]

**15** A curve C is given by the parametic equations $x = t^2$, $y = 2t$.

**(i)** Find the cartesian equation of the curve.

**(ii)** Find $\frac{dy}{dx}$ in terms of $t$. Hence, or otherwise, show that $\frac{dy}{dx} = \frac{y}{2x}$ at any point on the curve.

**(iii)** Another curve D has gradient given by $\frac{dy}{dx} = -\frac{2x}{y}$. Show that, at any point where C and D intersect the two curves are perpendicular.

**(iv)** Solve the differential equation $\frac{dy}{dx} = -\frac{2x}{y}$, and hence find the equation of D given that $y = 2$ when $x = 0$.

**(v)** Draw on the same axes a sketch showing the curves C and D.

[MEI]

**16 (i)** Express $\frac{1}{(3x-1)x}$ in partial fractions.

A model for the way in which a population of animals in a closed environment varies with time is given, for $P > \frac{1}{3}$, by

$$\frac{dP}{dt} = \tfrac{1}{2}(3P^2 - P)\sin t$$

where $P$ is the size of the population in thousands at time $t$.

**(ii)** Given that $P = \frac{1}{2}$ when $t = 0$, use the method of separation of variables to show that

$$\ln\left(\frac{3P-1}{P}\right) = \tfrac{1}{2}(1 - \cos t).$$

**(iii)** Calculate the smallest positive value of $t$ for which $P = 1$.

**(iv)** Rearrange the equation at the end of part (ii) to show that

$$P = \frac{1}{3 - e^{\frac{1}{2}(1-\cos t)}}.$$

Hence find the two values between which the number of animals in the population oscillates.

[MEI]

**17 (i)** Use integration by parts to show that

$$\int \ln x \, dx = x \ln x - x + c.$$

**(ii)** Differentiate $\ln(\sin x)$ with respect to $x$, for $0 < x < \frac{\pi}{2}$.

Hence write down $\int \cot x \, dx$, for $0 < x < \frac{\pi}{2}$.

**(iii)** For $x > 0$ and $0 < y < \frac{\pi}{2}$, the variables $y$ and $x$ are connected by the differential equation

$$\frac{dy}{dx} = \frac{\ln x}{\cot y},$$

and $y = \frac{\pi}{6}$ when $x = e$.

Find the value of $y$ when $x = 1$, giving your answer correct to 3 significant figures. Use the differential equation to show that this value of $y$ is a stationary value, and determine its nature.

[MEI]

**18** A wind is blowing offshore and so the waves become larger the further from the shore you travel. At the water's edge the waves have zero height. Three models are considered for the rate of increase in wave height $h$ with respect to distance $s$ from the shore.

**(a)** Rate of increase of $h$ with respect to $s$ is proportional to $s$.

**(b)** Rate of increase of $h$ with respect to $s$ is inversely proportional to $(s + 5)$.

**(c)** Rate of increase of $h$ with respect to $s$ is proportional to $e^{-cs}$, where $c$ is a positive constant.

**(i)** For each of these models, form and solve a differential equation.

**(ii)** For each model, sketch the graph of $h$ against $s$.

**(iii)** Discuss which of the models is the most realistic. In particular, consider the behaviour for large values of $s$.

## INVESTIGATION

Investigate the tea cooling problem introduced on page 179. You will need to make some assumptions about the initial temperature of the tea and the temperature of the room.

What difference would it make if you were to add some cold milk to the tea and then leave it to cool?

Would it be better to allow the tea to cool first before adding the milk?

## KEY POINTS

**1** A differential equation is an equation involving derivatives such as

$$\frac{dy}{dx} \quad \frac{d^2y}{dx^2}.$$

**2** A first-order differential equation involves a first derivative only.

**3** Some first-order differential equations may be solved by separating the variables.

**4** A general solution is one in which the constant of integration is left in the solution, and a particular solution is one in which additional information is used to calculate the constant of integration.

**5** A general solution may be represented by a family of curves, a particular solution by a particular member of that family.

# Answers



## Chapter 1

### Activity (Page 5)

For $|x| < 1$ the sum of the geometric series is $\frac{1}{1+x}$ which is the same as $(1 + x)^{-1}$.

### ❓ (Page 7)

$$\begin{aligned}\sqrt{101} &= \sqrt{100 \times 1.01}\\ &= 10\sqrt{1.01}\\ &= 10(1 + 0.01)^{\frac{1}{2}}\\ &= 10\left[1 + \tfrac{1}{2}(0.01) + \frac{(\frac{1}{2})(-\frac{1}{2})}{2!}(0.01)^2 + \ldots\right]\\ &= 10.050 \text{ (3 d.p.)}\end{aligned}$$

### ❓ (Page 9)

$\sqrt{x-1}$ is only defined for $x > 1$.
A possible rearrangement is $\sqrt{x(1-\frac{1}{x})} = \sqrt{x}(1-\frac{1}{x})^{\frac{1}{2}}$
Since $x > 1 \Rightarrow 0 < \frac{1}{x} < 1$ the binomial expansion could be used but the resulting expansion would not be a series of positive powers of $x$.

### Exercise 1A (Page 9)

**1 (i)** $1 - 2x + 3x^2$ **(ii)** $|x| < 1$ **(iii)** 0.43%

**2 (i)** $1 - 2x + 4x^2$ **(ii)** $|x| < \frac{1}{2}$ **(iii)** 0.8%

**3 (i)** $1 - \frac{x^2}{2} - \frac{x^4}{8}$ **(ii)** $|x| < 1$ **(iii)** 0.000 006 3%

**4 (i)** $1 + 4x + 8x^2$ **(ii)** $|x| < \frac{1}{2}$ **(iii)** 1.3%

**5 (i)** $\frac{1}{3} - \frac{x}{9} + \frac{x^2}{27}$ **(ii)** $|x| < 3$ **(iii)** 0.0037%

**6 (i)** $2 - \frac{7x}{4} - \frac{17x^2}{64}$ **(ii)** $|x| < 4$ **(iii)** 0.000 95%

**7 (i)** $-\frac{2}{3} - \frac{5x}{9} - \frac{5x^2}{27}$ **(ii)** $|x| < 3$ **(iii)** 0.0088%

**8 (i)** $\frac{1}{2} - \frac{3x}{16} + \frac{27x^2}{256}$ **(ii)** $|x| < \frac{4}{3}$ **(iii)** 0.013%

**9 (i)** $1 + 6x + 20x^2$ **(ii)** $|x| < \frac{1}{2}$ **(iii)** 4%

**10 (i)** $1 + 2x^2 + 2x^4$ **(ii)** $|x| < 1$ **(iii)** 0.000 20%

**11 (i)** $1 + \frac{2x^2}{3} - \frac{4x^4}{9}$ **(ii)** $|x| < \frac{1}{\sqrt{2}}$ **(iii)** 0.000 048%

**12 (i)** $1 - 3x + 7x^2$ **(ii)** $|x| < \frac{1}{2}$ **(iii)** 1.64%

**13 (i)** $1 + 3x + 3x^2 + x^3$
**(ii)** $1 + 4x + 10x^2 + 20x^3$ for $|x| < 1$
**(iii)** $a = 25$, $b = 63$

**14 (i)** $16 - 32x + 24x^2 - 8x^3 + x^4$
**(ii)** $1 - 6x + 24x^2 - 80x^3$ for $|x| < \frac{1}{2}$
**(iii)** $a = -128$, $b = 600$

**15 (i)** $1 + x + x^2 + x^3$ for $|x| < 1$
**(ii)** $1 - 4x + 12x^2 - 32x^3$ for $|x| < \frac{1}{2}$
**(iii)** $1 - 3x + 9x^2 - 23x^3$ for $|x| < \frac{1}{2}$

**16 (ii)** $1 + \frac{x}{8} + \frac{3x^2}{128}$ for $|x| < 4$
**(iii)** $1 + \frac{9x}{8} + \frac{19x^2}{128}$

**17 (i)** $1 - y + y^2 - y^3 \ldots$
**(ii)** $1 - \frac{2}{x} + \frac{4}{x^2} - \frac{8}{x^3}$
**(iv)** $\frac{x}{2} - \frac{x^2}{4} + \frac{x^3}{8} - \frac{x^4}{16}$
**(v)** $x < -2$ or $x > 2$; $-2 < x < 2$;
no overlap in range of validity.

### Exercise 1B (Page 13)

**1** $\frac{2a^2}{3b^3}$ **2** $\frac{1}{9y}$ **3** $\frac{x+3}{x-6}$ **4** $\frac{x+3}{x+1}$

**5** $\frac{2x-5}{2x+5}$ **6** $\frac{3(a+4)}{20}$ **7** $\frac{x(2x+3)}{(x+1)}$ **8** $\frac{2}{5(p-2)}$

**9** $\frac{a-b}{2a-b}$ **10** $\frac{(x+4)(x-1)}{x(x+3)}$ **11** $\frac{9}{20x}$ **12** $\frac{x-3}{12}$

**13** $\frac{a^2+1}{a^2-1}$ **14** $\frac{5x-13}{(x-3)(x-2)}$ **15** $\frac{2}{(x+2)(x-2)}$

**16** $\frac{2p^2}{(p^2-1)(p^2+1)}$ **17** $\frac{a^2-a+2}{(a+1)(a^2+1)}$ **18** $\frac{-2(y^2+4y+8)}{(y+2)^2(y+4)}$

**19** $\frac{x^2+x+1}{x+1}$ **20** $-\frac{(3b+1)}{(b+1)^2}$ **21** $\frac{13x-5}{6(x-1)(x+1)}$

**22** $\frac{4(3-x)}{5(x+2)^2}$ **23** $\frac{3a-4}{(a+2)(2a-3)}$ **24** $\frac{3x^2-4}{x(x-2)(x+2)}$

### Exercise 1C (Page 16)

**1 (i)** 84 **(ii)** 4 **(iii)** –2
**(iv)** 5.24 or 0.76 **(v)** 3 or $\frac{1}{3}$ **(vi)** 0 or 3
**(vii)** 1.71 or 0.29

**2 (i)** $\frac{600}{x}$ **(ii)** $\frac{600}{x-1}$ **(iii)** $x^2 - x - 600 = 0,\ x = 25$

**3 (i)** $\frac{270}{x}, \frac{270}{x-10}$ **(ii)** $x^2 - 10x - 9000 = 0,\ x = 100$
**(iii)** Arrive 1 pm

**4** Cost = £16, 16 staff left

**5** 12 thick slices

**6 (i)** 1.714 ohms **(ii)** 4 ohms
**(iii)** Equivalent to half

### ? (Page 20)

The identity is true for all values of $x$. Once a particular value of $x$ is substituted you have an equation. Equating constant terms is equivalent to substituting $x = 0$.

### Exercise 1D (Page 21)

**1** $\frac{1}{(x-2)} - \frac{1}{(x+3)}$ **2** $\frac{1}{x} - \frac{1}{(x+1)}$ **3** $\frac{2}{(x-4)} - \frac{2}{(x-1)}$

**4** $\frac{2}{(x-1)} - \frac{1}{(x+2)}$ **5** $\frac{1}{(x+1)} + \frac{1}{(2x-1)}$ **6** $\frac{2}{(x-2)} - \frac{2}{x}$

**7** $\frac{1}{(x-1)} - \frac{3}{(3x-1)}$ **8** $\frac{3}{5(x-4)} + \frac{2}{5(x+1)}$

**9** $\frac{5}{(2x-1)} - \frac{2}{x}$ **10** $\frac{2}{(2x-3)} - \frac{1}{(x+2)}$

**11** $\frac{8}{13(2x-5)} + \frac{9}{13(x+4)}$ **12** $\frac{19}{24(3x-2)} - \frac{11}{24(3x+2)}$

### Exercise 1E (Page 23)

**1** $\frac{9}{(1-3x)} - \frac{3}{(1-x)} - \frac{2}{(1-x)^2}$ **2** $\frac{4}{(2x-1)} - \frac{2x}{(x^2+1)}$

**3** $\frac{1}{(x-1)^2} - \frac{1}{(x-1)} + \frac{1}{(x+2)}$ **4** $\frac{5}{8(x-2)} + \frac{6-5x}{8(x^2+4)}$

**5** $\frac{5-2x}{(2x^2-3)} + \frac{2}{x+2}$ **6** $\frac{2}{x} - \frac{1}{x^2} - \frac{3}{(2x+1)}$

**7** $\frac{10x}{(3x^2-1)} - \frac{3}{x}$

**8** $\frac{1}{(2x^2+1)} + \frac{1}{(x+1)}$

**9** $\frac{8}{(2x-1)} - \frac{4}{(2x-1)^2} - \frac{3}{x}$

**10** $A = 1,\quad B = 0,\quad C = 1$

**11** $A = 1,\quad B = 0,\quad C = -4$

### Exercise 1F (Page 25)

**1** $4 + 20x + 72x^2$ **2** $-4 - 10x - 16x^2$

**3** $\frac{5}{2} + \frac{11x}{4} + \frac{33x^2}{8}$ **4** $-\frac{1}{8} - \frac{5x}{16} - \frac{x^2}{8}$

**5 (i)** $\frac{2}{(2x-1)} - \frac{3}{(x+2)}$
**(ii)** $1 + 2x + 4x^2 \ldots\ a = 1,\ b = 2,\ c = 4$, for $|x| < \frac{1}{2}$
**(iii)** $\frac{1}{2} - \frac{x}{4} + \frac{x^2}{8}$ for $|x| < 2$
**(iv)** $-\frac{7}{2} - \frac{13x}{4} - \frac{67x^2}{8}$; 0.505%

## Chapter 2

### Activity (Page 28)

$y = \sin(\theta + 60°)$ is obtained from $y = \sin\theta$ by a translation $\begin{pmatrix}-60°\\0\end{pmatrix}$

$y = \cos(\theta - 60°)$ is obtained from $y = \sin\theta$ by a translation $\begin{pmatrix}60°\\0\end{pmatrix}$

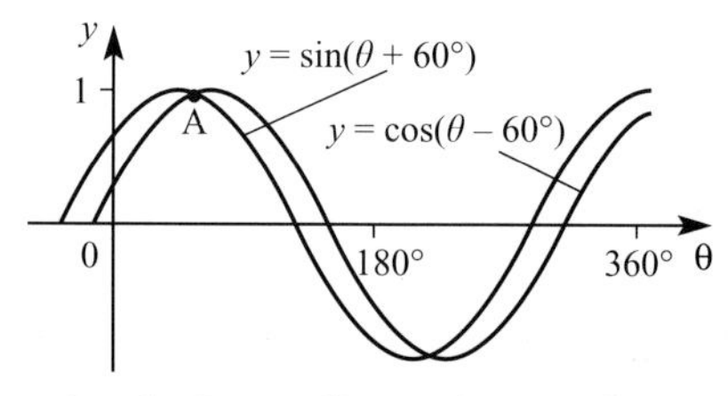


It appears that the $\theta$ co-ordinate of A is midway between the two maxima (30°, 1) and (60°, 1).

Checking: $\theta = 45° \rightarrow \sin(\theta + 60°) = 0.966$
$\cos(\theta - 60°) = 0.966.$

If 60° is replaced by 35°, using the trace function on a graphics calculator would enable the solutions to be found.

### Activity (Page 29)

**(i)** $\sin(\theta + \phi) = \sin\theta\cos\phi + \cos\theta\sin\phi$
$\Rightarrow \sin[(90° - \theta) + \phi] = \sin(90° - \theta)\cos\phi + \cos(90° - \theta)\sin\phi$
$\Rightarrow \sin[90° - (\theta - \phi)] = \cos\theta\cos\phi + \sin\theta\sin\phi$
$\Rightarrow \cos(\theta - \phi) = \cos\theta\cos\phi + \sin\theta\sin\phi$

**(ii)** $\Rightarrow \cos[\theta - (-\phi)] = \cos\theta\cos(-\phi) + \sin\theta\sin(-\phi)$
$\cos(\theta + \phi) = \cos\theta\cos\phi - \sin\theta\sin\phi$

**(iii)**
$$\tan(\theta + \phi) = \frac{\sin(\theta + \phi)}{\cos(\theta + \phi)} = \frac{\sin\theta\cos\phi + \cos\theta\sin\phi}{\cos\theta\cos\phi - \sin\theta\sin\phi} = \frac{\frac{\sin\theta\cos\phi}{\cos\theta\cos\phi} + \frac{\cos\theta\sin\phi}{\cos\theta\cos\phi}}{\frac{\cos\theta\cos\phi}{\cos\theta\cos\phi} - \frac{\sin\theta\sin\phi}{\cos\theta\cos\phi}} = \frac{\tan\theta + \tan\phi}{1 - \tan\theta\tan\phi}$$

**(iv)** $\tan[\theta + (-\phi)] = \dfrac{\tan\theta + \tan(-\phi)}{1 - \tan\theta\tan(-\phi)}$

$\tan(\theta - \phi) = \dfrac{\tan\theta - \tan\phi}{1 + \tan\theta\tan\phi}$

## Exercise 2A (Page 31)

**1 (i)** $\dfrac{\sqrt{3}}{2\sqrt{2}} + \dfrac{1}{2\sqrt{2}}$ **(ii)** $-\dfrac{1}{\sqrt{2}}$

**(iii)** $\dfrac{\sqrt{3}-1}{\sqrt{3}+1}$ **(iv)** $\dfrac{\sqrt{3}+1}{\sqrt{3}-1}$

**2 (i)** $\frac{1}{\sqrt{2}}(\sin\theta + \cos\theta)$ **(ii)** $\frac{1}{2}(\sqrt{3}\cos\theta + \sin\theta)$

**(iii)** $\frac{1}{2}(\sqrt{3}\cos\theta - \sin\theta)$ **(iv)** $\frac{1}{\sqrt{2}}(\cos 2\theta - \sin 2\theta)$

**(v)** $\dfrac{\tan\theta + 1}{1 - \tan\theta}$ **(vi)** $\dfrac{\tan\theta - 1}{1 + \tan\theta}$

**3 (i)** $\sin\theta$ **(ii)** $\cos 4\phi$ **(iii)** 0 **(iv)** $\cos 2\theta$

**4 (i)** 15° **(ii)** 157.5° **(iii)** 0° or 180° **(iv)** 111.7°

**(v)** 165°

**5 (i)** $\frac{\pi}{8}$ **(ii)** 2.79 radians

**6 (i)** $\dfrac{1}{\sqrt{5}}$ **(ii)** $\sin\beta = \dfrac{3}{5}, \cos\beta = \dfrac{4}{5}$

## Exercise 2B (Page 36)

**1 (i)** 14.5°, 90°, 165.5°, 270°

**(ii)** 0°, 35.3°, 144.7°, 180°, 215.3°, 324.7°, 360°

**(iii)** 90°, 210°, 330°

**(iv)** 30°, 150°, 210°, 330°

**(v)** 0°, 138.6°, 221.4°, 360°

**2 (i)** $-\pi, 0, \pi$ **(ii)** $-\pi, 0, \pi$ **(iii)** $\frac{-2\pi}{3}, 0, \frac{2\pi}{3}$

**(iv)** $\frac{-3\pi}{4}, \frac{-\pi}{4}, \frac{\pi}{4}, \frac{3\pi}{4}$, **(v)** $\frac{-11\pi}{12}, \frac{-3\pi}{4}, \frac{-7\pi}{12}, \frac{-\pi}{4}, \frac{\pi}{12}, \frac{\pi}{4}, \frac{5\pi}{12}, \frac{3\pi}{4}$

**3** $3\sin\theta - 4\sin^3\theta$, $\theta = 0, \frac{\pi}{4}, \frac{3\pi}{4}, \pi, \frac{5\pi}{4}, \frac{7\pi}{4}, 2\pi$

**4** 51°, 309°

**5** $\cot\theta$

**6** $\dfrac{\tan\theta(3 - \tan^2\theta)}{1 - 3\tan^2\theta}$

**8 (ii)** 63.4°

**9 (i)**

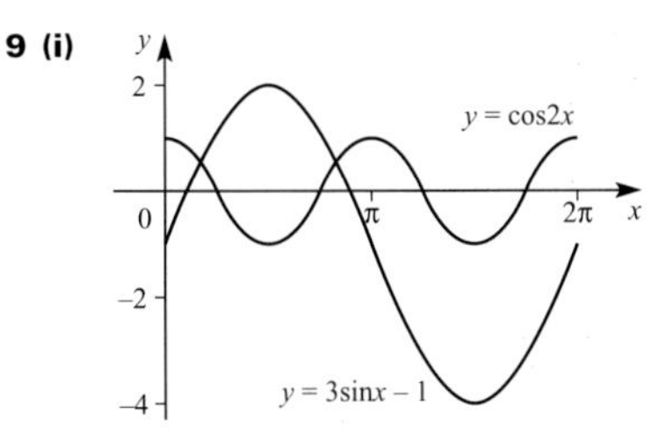


**(iii)** $x = \frac{\pi}{6}$ or $x = \frac{5\pi}{6}$

## Activity (Page 38)

$\cos(\theta + \phi) = \cos\theta\cos\phi - \sin\theta\sin\phi$ ①

$\cos(\theta - \phi) = \cos\theta\cos\phi + \sin\theta\sin\phi.$ ②

Adding ① and ②

$\cos(\theta + \phi) + \cos(\theta - \phi) = 2\cos\theta\cos\phi$

Let $\theta + \phi = \alpha$; $\theta - \phi = \beta$;

$\Rightarrow \cos\alpha + \cos\beta = 2\cos\left(\dfrac{\alpha+\beta}{2}\right)\cos\left(\dfrac{\alpha-\beta}{2}\right)$

Similarly, subtracting ② from ①

$\Rightarrow \cos(\theta + \phi) - \cos(\theta - \phi) = -2\sin\theta\sin\phi$

$\Rightarrow \cos\alpha - \cos\beta = -2\sin\left(\dfrac{\alpha+\beta}{2}\right)\sin\left(\dfrac{\alpha-\beta}{2}\right)$

## Exercise 2C (Page 40)

**1 (i)** $2\cos 3\theta\sin\theta$ **(ii)** $2\cos 3\theta\cos 2\theta$

**(iii)** $-2\sin 5\theta\sin 2\theta$ **(iv)** $\cos\theta$

**(v)** $\sqrt{2}\sin 3\theta$

**2** $2\cos 3\theta\cos\theta$

20°, 90°, 100°, 140°

**3** $\dfrac{\tan 4\theta}{\tan\theta}$

**4** $0, \frac{\pi}{4}, \frac{3\pi}{4}, \pi, \frac{5\pi}{4}, \frac{7\pi}{4}, 2\pi$

**5** $\cos(\theta + 43°)$

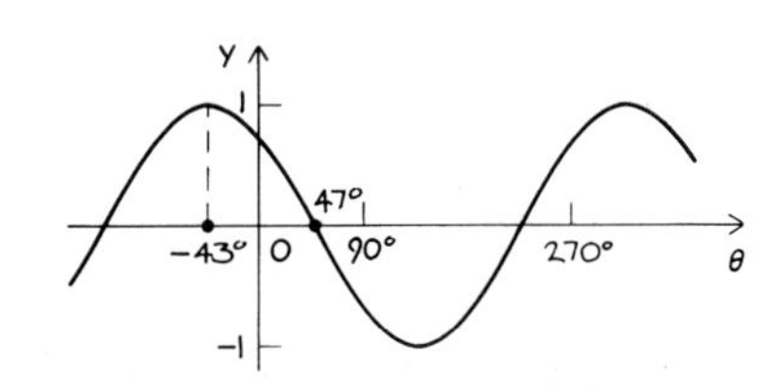


## Exercise 2D (Page 44)

**1 (i)** $\sqrt{2}\cos(\theta - 45°)$ **(ii)** $5\cos(\theta - 53.1°)$

**(iii)** $2\cos(\theta - 60°)$ **(iv)** $3\cos(\theta - 41.8°)$

**2 (i)** $\sqrt{2}\cos(\theta + \frac{\pi}{4})$ **(ii)** $2\cos(\theta + \frac{\pi}{6})$

**3 (i)** $\sqrt{5}\sin(\theta + 63.4°)$ **(ii)** $5\sin(\theta + 53.1°)$

**4 (i)** $\sqrt{2}\sin(\theta - \frac{\pi}{4})$ **(ii)** $2\sin(\theta - \frac{\pi}{6})$

**5 (i)** $2\cos(\theta - (-60°))$ **(ii)** $4\cos(\theta - (-45°))$

**(iii)** $2\cos(\theta - 30°)$ **(iv)** $13\cos(\theta - 22.6°)$

**(v)** $2\cos(\theta - 150°)$ **(vi)** $2\cos(\theta - 135°)$

**6 (i)** $13\cos(\theta + 67.4°)$ **(ii)** Max 13, min −13

**(iii)**

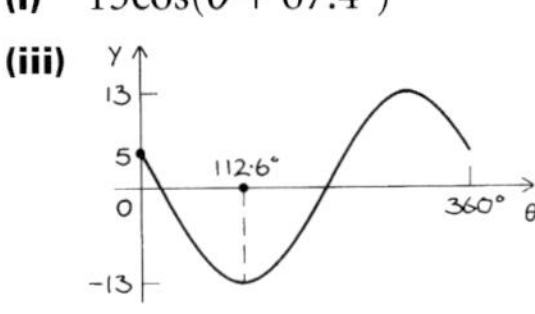


**(iv)** 4.7°, 220.5°

**7 (i)** $2\sqrt{3}\sin(\theta - \frac{\pi}{6})$

**(ii)** Max $2\sqrt{3}$, $\theta = \frac{2\pi}{3}$; min $-2\sqrt{3}$, $\theta = \frac{5\pi}{3}$

**(iii)**

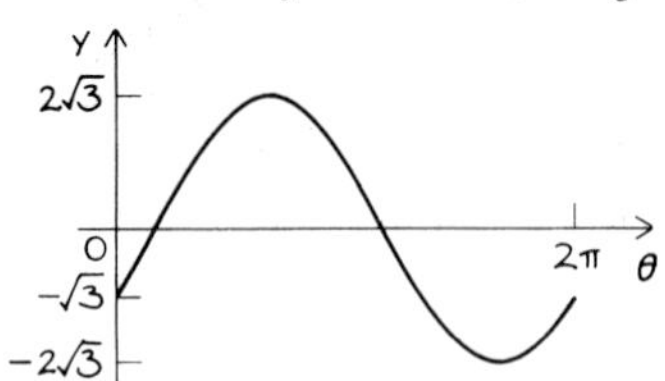


**(iv)** $\frac{\pi}{3}$, $\pi$

**8 (i)** $\sqrt{13}\sin(2\theta + 56.3°)$

**(ii)** Max $\sqrt{13}$, $\theta = 16.8°$; min $-\sqrt{13}$, $\theta = 106.8°$

**(iii)**

**(iv)** 53.8° 159.9°, 233.8°, 339.9°

**9 (i)** $\sqrt{3}\cos(\theta - 54.7°)$

**(ii)** Max $\sqrt{3}$, $\theta = 54.7°$; min $-\sqrt{3}$, $\theta = 234.7°$

**(iii)**

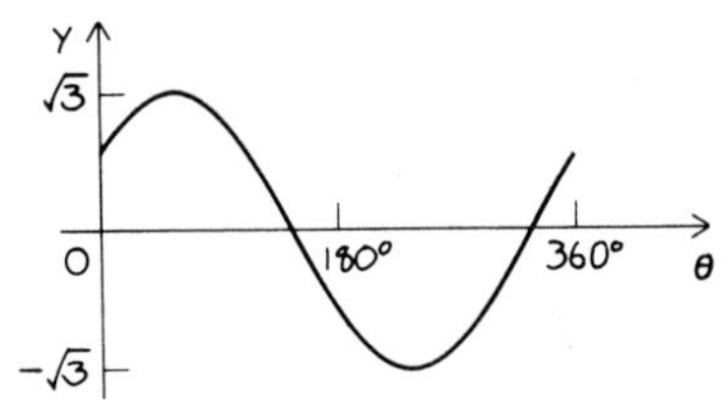


**(iv)** Max $\frac{1}{3-\sqrt{3}}$, $\theta = 234.7°$; min $\frac{1}{3+\sqrt{3}}$, $\theta = 54.7°$

**10 (ii)** 30.6° or 82.0°

**11 (i)** $\cos x\cos\alpha - \sin x\sin\alpha$ **(ii)** $r = \sqrt{29}$, $\alpha = 68.2°$

**(iii)** Max $\sqrt{29}$ when $x = 291.8°$, min $-\sqrt{29}$ when $x = 111.8°$

**(iv)** $x = 235.7°$ or $347.9°$

**12 (i)** $\sqrt{34}\cos(x + 30.96°)$ **(ii)** $x = 15.7°$ or $282.4°$

**(iii)** $x = 7.9°$ or $141.2°$ or $187.9°$ or $321.2°$

**13 (i)** $R = 10$, $\alpha = 53.13°$

**(ii)**

**(iii)** $x = 119.55°$ or $346.71°$

**(iv)** $\theta = 103.29°$ or $330.45°$

**14 (i)** $R = \sqrt{10}$, $\alpha = 18.43°$ **(ii)** $x = 90°$ or $306.9°$

**(iii)** $x = 90°$, $233.1°$ or $306.9°$

**(iv)** Part (iii) also contains solutions to $-3\cos x = 1 - \sin x$

**15 (i)** $c = \sqrt{a^2 + b^2}$ **(ii)** $\tan\alpha = \frac{b}{a}$

**(iii)** $\alpha = 36.87°$ **(iv)** $\theta = 103.29°$ or $330.45°$

## Exercise 2E (Page 48)

**1 (i)** $\sin 6\theta$ **(ii)** $\cos 6\theta$ **(iii)** 1

**(iv)** $\cos\theta$ **(v)** $\sin\theta$ **(vi)** $\frac{3}{2}\sin 2\theta$

**(vii)** $\cos\theta$ **(viii)** $-1$

**2 (i)** $1 - \sin 2x$ **(ii)** $\cos 2x$ **(iii)** $\frac{1}{2}(5\cos 2x - 1)$

**4 (i)** 4.4°, 95.6° **(ii)** 199.5°, 340.5°

**(iii)** $\frac{-\pi}{6}, \frac{\pi}{2}$ **(iv)** −15.9°, 164.1°

**(v)** $\frac{\pi}{6}, \frac{\pi}{2}, \frac{5\pi}{6}$ **(vi)** 20.8°, 122.3°

**(vii)** 76.0°, 135°

## ? (Page 51)

By the shape and symmetry of the graphs, in each case the maximum percentage error will occur for $\theta = 0.1$ radians.

$y = \sin\theta$: $\theta = 0.1$ rad
true value = 0.099 833
approximate value = 0.1
% error = 0.167%

$y = \tan\theta$: $\theta = 0.1$ rad
true value = 0.100 335
approximate value = 0.1
% error = 0.334%

$y = \cos\theta$: $\theta = 0.1$ rad
true value = 0.995 004
approximate value = 0.995
% error = 0.000 419%

## Activity (Page 51)

**(i)** When $\theta = 0$, $\frac{\cos\theta - \cos 2\theta}{\theta^2} = \frac{1-1}{0}$

$= \frac{0}{0}$ (undefined)

**(ii)**

| $\theta$ | 0.20 | 0.18 | 0.16 | 0.14 | 0.12 | 0.10 | 0.08 | 0.06 | 0.04 | 0.02 |
|---|---|---|---|---|---|---|---|---|---|---|
| $\frac{\cos\theta - \cos 2\theta}{\theta^2}$ | 1.475 | 1.480 | 1.484 | 1.488 | 1.491 | 1.494 | 1.496 | 1.498 | 1.499 | 1.500 |

## Exercise 2F (Page 53)

**1 (i)** 2 **(ii)** $1 - \sqrt{3}\theta - \frac{\theta^2}{2}$ **(iii)** $1 - \frac{5\theta^2}{2}$ **(iv)** $\frac{1}{2}$

**(v)** $-3\theta$ **(vi)** $\theta\sin\alpha + \theta^2\cos\alpha$

**2 (i)** $5\theta$ **(ii)** 5

**3 (i)** $\frac{\theta^2}{2}$ **(ii)** $\frac{1}{8}$

**4 (i)** $\frac{\sqrt{3}\theta^2}{2}$ **(ii)** $2\theta^2$ **(iii)** $\frac{\sqrt{3}}{4}$

**5 (i)** $8\theta^2$ **(ii)** $4\theta^2$ **(iii)** 2

**6 (i)** $\frac{1}{1+\theta}$ **(ii)** $1-\theta+\theta^2$ **(iii)** 0.03% and 0.13%

**7 (i)** $\sqrt{1+\theta}$ **(ii)** $1+\frac{1}{2}\theta-\frac{1}{8}\theta^2$
**(iii)** $\sqrt{1+\theta}$ since this has only used one approximation.
**(iv)** $\sqrt{1+\theta} = 1.048\,81$, $1+\frac{1}{2}\theta-\frac{1}{8}\theta^2 = 1.048\,75$, true value 1.048 73.
Errors due to the double approximation appear to have cancelled out to some extent, rather than compounding.

**8 (i)** $\frac{1}{1-\frac{1}{2}\theta^2}$ **(ii)** $1+\frac{1}{2}\theta^2$ **(iii)** 0.47 radians

**9 (ii)** $\angle BAE = 90° - \angle OAB$

**10 (i)** $x = -\frac{\pi}{8}$ or $\frac{3\pi}{8}$ **(ii)** $2x$ **(iii)** $2x+1-2x^2$
**(iv)** $x = -0.366$ or $1.366$ **(v)** Angles in **(i)** are not 'small'

**11 (i)** $\frac{2}{1-x} - \frac{2}{2-x}$; $k = \frac{3}{2}$ **(iii)** $\theta = \pm\, 0.2$

**12 (i)** $\sin x\cos(\delta x) + \cos x\sin(\delta x)$
**(ii)** $\sin x + (\delta x)\cos x - \frac{(\delta x)^2}{2}\sin x$
**(iii)** $\cos x - \frac{(\delta x)}{2}\sin x$ **(iv)** $\cos x$ **(v)** Derivative of $\sin x$

### Activity (Page 58)

General solution is $\theta = 2n\pi + \arcsin c$ or $\theta = (2n+1)\pi - \arcsin c$

i.e. even multiples of $\pi$ are followed by $+ \arcsin c$
odd multiples of $\pi$ are followed by $- \arcsin c$.

Now $(-1)^n = +1$ when $n$ is even and
$(-1)^n = -1$ when $n$ is odd

so $\theta = n\pi + (-1)^n \arcsin c$.

## Chapter 3

### Activity (Page 61)

**(i)**

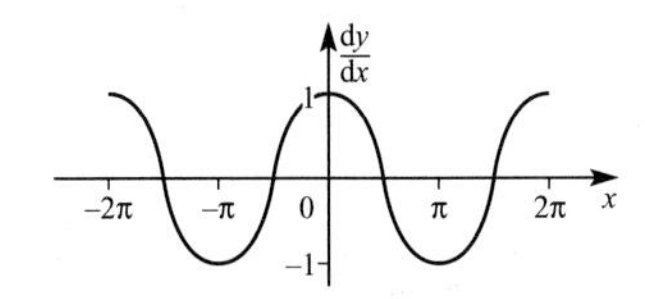


When $y = \sin x$ the graph of $\frac{dy}{dx}$ against $x$ looks like the graph of $\cos x$.

**(ii)**

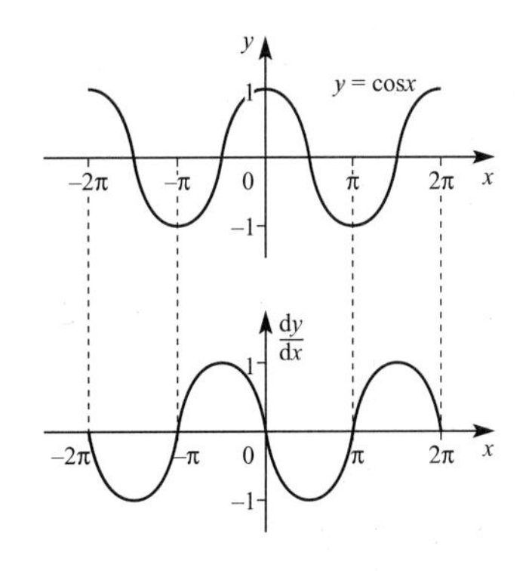


### Activity (Page 63)

$$\lim_{\delta x\to 0}\frac{\sin(x+\delta x)-\sin x}{\delta x} = \lim_{\delta x\to 0}\frac{2\sin\left(\frac{\delta x}{2}\right)\cos\left(x+\frac{\delta x}{2}\right)}{\delta x}$$

$$= \lim_{\delta x\to 0}\frac{2\times\left(\frac{\delta x}{2}\right)\times\cos\left(x+\frac{\delta x}{2}\right)}{\delta x}$$

$$= \lim_{\delta x\to 0}\cos\left(x+\frac{\delta x}{2}\right)$$

$$= \cos x$$

### Activity (Page 63)

**(i)** When $y = \cos x$

$$\frac{dy}{dx} = \lim_{\delta x\to 0}\frac{\cos(x+\delta x)-\cos x}{\delta x}$$

$$= \lim_{\delta x\to 0}\frac{[\cos x\cos\delta x - \sin x\sin\delta x] - \cos x}{\delta x}$$

$$= \lim_{\delta x\to 0}\frac{[\cos x\,(1-\frac{1}{2}(\delta x)^2) - (\sin x)\delta x] - \cos x}{\delta x}$$

$$= \lim_{\delta x\to 0}\frac{-\frac{1}{2}(\cos x)(\delta x)^2 - (\sin x)\delta x}{\delta x}$$

$$= \lim_{\delta x\to 0}\left(-\tfrac{1}{2}(\cos x)\delta x - \sin x\right)$$

$$= -\sin x$$

**(ii)** $y = \tan x = \frac{\sin x}{\cos x}$

$$\frac{dy}{dx} = \frac{\cos x(\cos x) - \sin x(-\sin x)}{\cos^2 x}$$

$$= \frac{\cos^2 x + \sin^2 x}{\cos^2 x}$$

$$= \frac{1}{\cos^2 x}$$

$$= \sec^2 x$$

## Exercise 3A (Page 66)

**1 (i)** $-2\sin x + \cos x$ **(ii)** $\sec^2 x$ **(iii)** $\cos x + \sin x$

**2 (i)** $x\sec^2 x + \tan x$ **(ii)** $\cos^2 x - \sin^2 x = \cos 2x$
**(iii)** $e^x(\sin x + \cos x)$

**3 (i)** $\dfrac{x\cos x - \sin x}{x^2}$ **(ii)** $\dfrac{e^x(\cos x + \sin x)}{\cos^2 x}$
**(iii)** $\dfrac{\sin x(1 - \sin x) - \cos x(x + \cos x)}{\sin^2 x}$

**4 (i)** $2x\sec^2(x^2 + 1)$ **(ii)** $-\sin 2x$ **(iii)** $\cot x$

**5 (i)** $-\dfrac{\sin x}{2\sqrt{\cos x}}$ **(ii)** $e^x(\tan x + \sec^2 x)$
**(iii)** $8x\cos 4x^2$ **(iv)** $-2\sin 2x e^{\cos 2x}$
**(v)** $\dfrac{1}{(1 + \cos x)}$ **(vi)** $\dfrac{1}{(\sin x\cos x)} = 2\text{cosec}2x$

**6 (i)** $\sec x\tan x$ **(ii)** $-\text{cosec}x\cot x$ **(iii)** $-\text{cosec}^2 x$

**7 (i)** $\cos x - x\sin x$ **(ii)** $-1$ **(iii)** $y = -x$
**(iv)** $y = x - 2\pi$

**8 (i)** $3\cos x\sin^2 x$
**(ii)** $(-\pi, 0)$ point of inflection, $(-\frac{1}{2}\pi, -1)$ min, $(0, 0)$ point of inflection, $(\frac{1}{2}\pi, 1)$ max, $(\pi, 0)$ point of inflection
**(iv)**

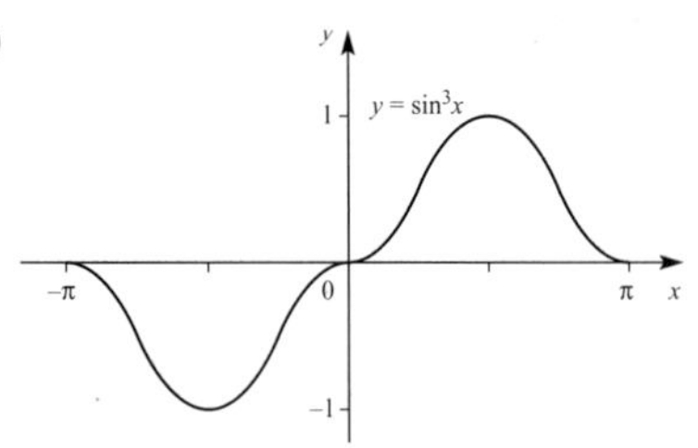


**9 (i)** $1 + 2\cos 2x$
**(ii)** $(\frac{\pi}{3}, \frac{\pi}{3} + \frac{\sqrt{3}}{2})$ max, $(\frac{2\pi}{3}, \frac{2\pi}{3} - \frac{\sqrt{3}}{2})$ min, $(\frac{4\pi}{3}, \frac{4\pi}{3} + \frac{\sqrt{3}}{2})$ max, $(\frac{5\pi}{3}, \frac{5\pi}{3} - \frac{\sqrt{3}}{2})$ min
**(iii)**

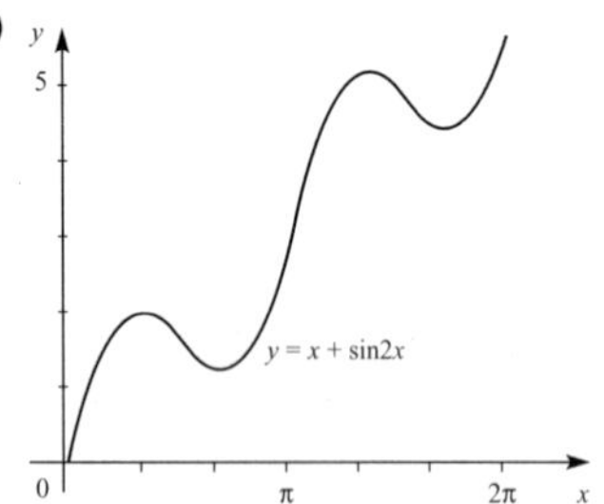


**11 (i)** $e^{-x}(\cos x - \sin x)$ **(iii)** $(0.79, 0.32)$, $(-2.4, -7.5)$

**12 (i) (b)** $2\cos 2x$ **(ii) (a)** $1 - \frac{1}{2}\sin^2 2x$

**13 (i)** $2e^{-x}\cos 2x - e^{-x}\sin 2x$ **(iii)** $x = 0.55$ or $2.12$
**(iv)** $r = \sqrt{5}$, $\alpha = 0.46$

**14** $e^{-2x}\sec^2 x - 2e^{-2x}\tan x$

## ❓ (Page 69)

**(i)** Interchange $x$ and $y$ to reflect in the line $y = x$.
**(ii)** The function is one-to-many.

## Exercise 3B (Page 72)

**1 (i)** $4y^3\dfrac{dy}{dx}$ **(ii)** $2x + 3y^2\dfrac{dy}{dx}$ **(iii)** $x\dfrac{dy}{dx} + y + 1 + \dfrac{dy}{dx}$
**(iv)** $-\sin y\dfrac{dy}{dx}$ **(v)** $e^{(y+2)}\dfrac{dy}{dx}$ **(vi)** $y^3 + 3xy^2\dfrac{dy}{dx}$
**(vii)** $4xy^5 + 10x^2y^4\dfrac{dy}{dx}$ **(viii)** $1 + \dfrac{1}{y}\dfrac{dy}{dx}$
**(ix)** $xe^y\dfrac{dy}{dx} + e^y + \sin y\dfrac{dy}{dx}$
**(x)** $\dfrac{x^2}{y}\dfrac{dy}{dx} + 2x\ln y$ **(xi)** $e^{\sin y} + x\cos y e^{\sin y}\dfrac{dy}{dx}$
**(xii)** $\tan y + (x\sec^2 y)\dfrac{dy}{dx} - (\tan x)\dfrac{dy}{dx} - y\sec^2 x$

**2** $\frac{1}{5}$

**3** 0

**4 (i)** $-1$ **(ii)** $x + y - 2 = 0$

**5** $(1, -2)$ and $(-1, 2)$

**6 (ii)** Max $(4, 8)$, min $(-4, -8)$

**7 (i)** $\dfrac{y + 4}{6 - x}$ **(ii)** $x - 2y - 11 = 0$
**(iii)** $(2, -4\frac{1}{2})$
**(iv)**

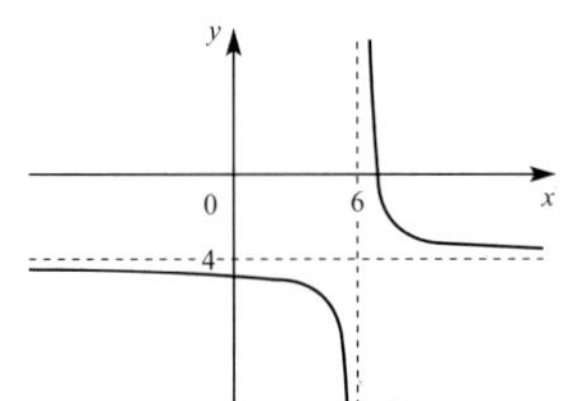


Asymptotes $x = 6$, $y = -4$

**8 (i)** $\ln y = x\ln x$
**(ii)** $\dfrac{1}{y}\dfrac{dy}{dx} = 1 + \ln x$
**(iii)** $(0.368, 0.692)$
**(iv)**

## Exercise 3C (Page 77)

**1 (i)** $-\cos x - 2\sin x + c$ **(ii)** $3\sin x - 2\cos x + c$
**(iii)** $-5\cos x + 4\sin x + c$

**2 (i)** $\frac{1}{3}\sin 3x + c$ **(ii)** $\cos(1 - x) + c$
**(iii)** $-\frac{1}{4}\cos^4 x + c$ **(iv)** $\ln|2 - \cos x| + c$
**(v)** $-\ln|\cos x| + c$ **(vi)** $-\frac{1}{6}(\cos 2x + 1)^3 + c$

**3 (i)** $-\cos(x^2) + c$ **(ii)** $e^{\sin x} + c$
**(iii)** $\frac{1}{2}\tan^2 x + c$ or $\frac{1}{2}\sec^2 x + c'$
**(iv)** $\dfrac{-1}{(\sin x)} + c$

**4 (i)** 1 **(ii)** $\frac{1}{16}$ **(iii)** 1
**(iv)** $e - 1$ **(v)** $\ln 2$

**5 (i)**

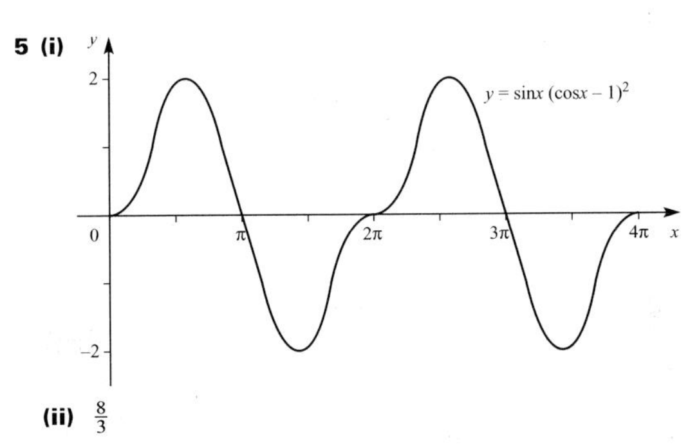


**(ii)** $\frac{8}{3}$

**6 (i)** $\frac{1}{4}\sin 2x + \frac{1}{2}x + c$ **(ii)** $-\cos x + \frac{1}{3}\cos^3 x + c$
**(iii)** $\frac{3}{8}x - \frac{1}{4}\sin 2x + \frac{1}{32}\sin 4x + c$
**(iv)** $\sin x - \frac{2}{3}\sin^3 x + \frac{1}{5}\sin^5 x + c$

## Activity (Page 79)

**(a) (i)** $\dfrac{d}{dx}(x\cos x) = -x\sin x + \cos x$

**(ii)** $\Rightarrow x\cos x = \int -x\sin x\,dx + \int \cos x\,dx$

$\Rightarrow \int x\sin x\,dx = -x\cos x + \int \cos x\,dx$

**(iii)** $\Rightarrow \int x\sin x\,dx = -x\cos x + \sin x + c$

**(b) (i)** $\dfrac{d}{dx}(xe^{2x}) = x \times 2e^{2x} + e^{2x}$

**(ii)** $\Rightarrow xe^{2x} = \int 2xe^{2x}\,dx + \int e^{2x}\,dx$

$\Rightarrow \int 2xe^{2x}\,dx = xe^{2x} - \int e^{2x}\,dx$

**(iii)** $\Rightarrow \int 2xe^{2x}\,dx = xe^{2x} - \frac{1}{2}e^{2x} + c$

## ? (Page 79)

Each of the integrals in the previous activity is of the form $\int x\dfrac{dv}{dx}\,dx$ and is found by starting with the product $xv$.

## Exercise 3D (Page 83)

**1 (i)** $u = x, \dfrac{dv}{dx} = e^x$ **(ii)** $xe^x - e^x + c$

**2 (i)** $u = x, \dfrac{dv}{dx} = \cos 3x$ **(ii)** $\frac{1}{3}x\sin 3x + \frac{1}{9}\cos 3x + c$

**3 (i)** $u = 2x + 1, \dfrac{dv}{dx} = \cos x$ **(ii)** $(2x + 1)\sin x + 2\cos x + c$

**4 (i)** $u = x, \dfrac{dv}{dx} = e^{-2x}$ **(ii)** $-\frac{1}{2}xe^{-2x} - \frac{1}{4}e^{-2x} + c$

**5 (i)** $u = x, \dfrac{dv}{dx} = e^{-x}$ **(ii)** $-xe^{-x} - e^{-x} + c$

**6 (i)** $u = x, \dfrac{dv}{dx} = \sin 2x$ **(ii)** $-\frac{1}{2}x\cos 2x + \frac{1}{4}\sin 2x + c$

**7** $\frac{1}{4}x^4\ln x - \frac{1}{16}x^4 + c$

**8** $x^2e^x - 2xe^x + 2e^x + c$

**9** $(2 - x)^2\sin x - 2(2 - x)\cos x - 2\sin x + c$

**10** $\frac{1}{3}x^3\ln 2x - \frac{1}{9}x^3 + c$

**11** $\frac{2}{15}(1 + x)^{\frac{3}{2}}(3x - 2) + c$

**12** $\frac{1}{15}(x - 2)^5(5x + 2) + c$

**13 (i)** $x\ln x - x + c$ **(ii)** $x\ln 3x - x + c$

## Exercise 3E (Page 86)

**1** $\frac{2}{9}e^3 + \frac{1}{9}$

**2** $-2$

**3** $2e^2$

**4** $3\ln 2 - 1$

**5** $\frac{1}{8}\pi^2 - \frac{1}{2}$

**6** $\frac{64}{3}\ln 4 - 7$

**7 (i)** $(2, 0), (0, 2)$

**(ii)**

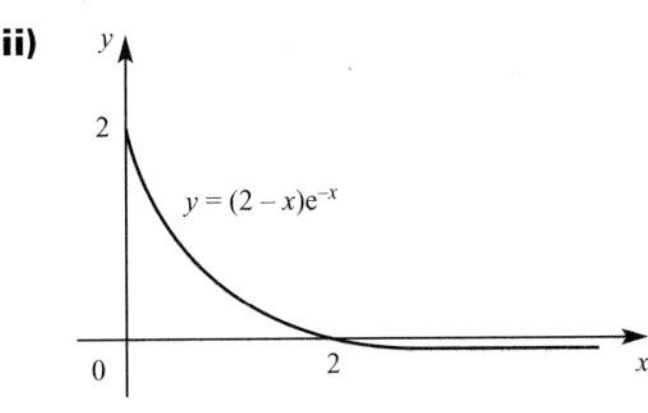


**(iii)** $e^{-2} + 1$

**8 (i)**

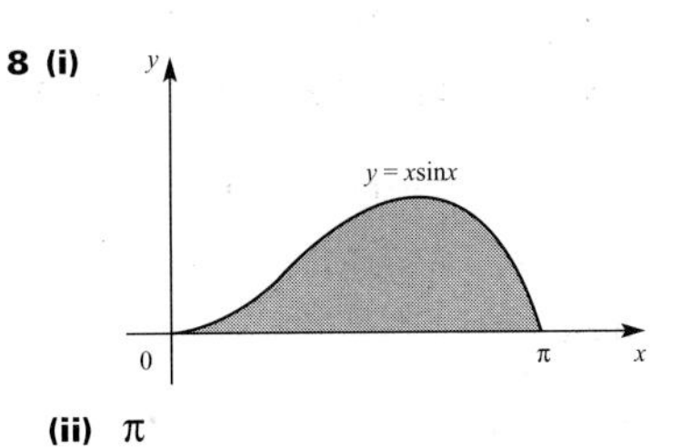


**(ii)** $\pi$

**9** $5\ln 5 - 4$

**10** $\frac{1}{2}\pi^2 - 4$

**11** $-\frac{4}{15}$ so area $= \frac{4}{15}$ square units

**12** $x = 0.5$; area $= 0.134$ square units

**13 (ii)** $1 + x + \dfrac{3x^2}{2} + \dfrac{5x^3}{2}$ for $|x| < \frac{1}{2}$

**(iii)** 0.005 16

**14 (i)** $\dfrac{1}{k}x\sin kx + \dfrac{1}{k^2}\cos kx + c$ **(ii)** $\cos 2x - \cos 8x$

**15 (i)** $\frac{1}{2}\theta\sin 2\theta + \frac{1}{4}\cos 2\theta + c$

**(ii)** $1 - 6x + 24x^2 - 80x^3$ for $|x| < \frac{1}{2}$

**(iii)** $a = 1, b = 6$; 0.005 15

**16** Curve is below trapezia

**17** $I_3 = 6 - 16e^{-1}$

**? (Page 88)**

Substitution using $u = x^2 - 1$ needs $2x$ in the numerator.
Not a product, not suitable for integration by parts.

**Exercise 3F (Page 91)**

**1 (i)** $\ln\left|\dfrac{3x-2}{1-x}\right| + c$ **(ii)** $\dfrac{1}{1-x} + \ln\left|\dfrac{x-1}{2x+3}\right| + c$

**(iii)** $\ln\left|\dfrac{x-1}{\sqrt{x^2+1}}\right| + c$ **(iv)** $\ln\left|\dfrac{(x-1)^2}{\sqrt{2x+1}}\right| + c$

**(v)** $\ln\left|\dfrac{x}{1-x}\right| - \dfrac{1}{x} + c$ **(vi)** $\frac{1}{2}\ln\left|\dfrac{x+1}{x+3}\right| + c$

**(vii)** $\ln\left|\dfrac{\sqrt{x^2+4}}{x+2}\right| + c$ **(viii)** $\ln\left|\dfrac{2x+1}{x+2}\right| + \dfrac{1}{2(2x+1)} + c$

**2** $-\dfrac{x}{x^2+4} + \dfrac{1}{x-3}$, $\ln\left(\dfrac{\sqrt{2}}{6}\right)$

**3** $\dfrac{1}{x^2} - \dfrac{2}{x} + \dfrac{4}{2x+1}$

**4 (i) (a)** $\dfrac{2}{1-2x} + \dfrac{1}{1+x}$ **(b)** $\ln(\frac{11}{8}) = 0.318\,45$

**(ii) (a)** $3 + 3x + 9x^2 + \ldots$ **(b)** 0.318 00 **(c)** 0.14%

**5 (i)** $A = 1, B = 3, C = -2$ **(ii)** $2 + \ln(\frac{125}{3}) = 5.73$

**6 (i)** $\frac{1}{2}xe^{2x} - \frac{1}{4}e^{2x} + c$ **(ii)** $\dfrac{9\pi - 2}{24}$

**(iii)** $A = 8, C = 1$

**7 (i)** $\dfrac{2}{1+2x} - \dfrac{(x+1)}{1+x^2}$ **(ii)** $|x| < \frac{1}{2}$ **(iii)** 0.078

**8 (i)** $B = 1, C = 16$ **(ii)** $\frac{33}{2}\ln 2$

**(iii)** $8 + 5x + 2x^2 + \dfrac{x^4}{2}$ for $|x| < 1$

**? (Page 93)**

**(i)** Partial fractions
**(ii)** Substitution: $u = x^2 + 2x - 3$
**(iii)** Integration by parts
**(iv)** Substitution: $u = x^2$
**(v)** Write $\sin^2 x = \frac{1}{2}(1 - \cos 2x)$
**(vi)** Substitution: $u = \sin x$

**Activity (Page 94)**

**(i)** $\displaystyle\int \frac{x-5}{x^2+2x-3}\,dx = \int \frac{2}{(x+3)}\,dx - \int \frac{1}{(x-1)}\,dx$
$= 2\ln|x+3| - \ln|x-1| + c$

**(ii)** $\displaystyle\int \frac{x+1}{x^2+2x-3}\,dx = \frac{1}{2}\int \frac{2x+2}{x^2+2x-3}\,dx$
$= \frac{1}{2}\ln|x^2+2x-3| + c$

**(iii)** $\displaystyle\int xe^x\,dx = xe^x - \int e^x\,dx$
$= xe^x - e^x + c$

**(iv)** $\displaystyle\int xe^{x^2}\,dx = \int \frac{1}{2}e^u\,du$ where $u = x^2$
$= \frac{1}{2}e^u + c$
$= \frac{1}{2}e^{x^2} + c$

**(v)** $\displaystyle\int \sin^2 x\,dx = \int \frac{1}{2}(1 - \cos 2x)\,dx$
$= \frac{x}{2} - \frac{1}{4}\sin 2x + c$

**(vi)** $\displaystyle\int \cos x\sin^2 x\,dx = \int u^2\,du$ where $u = \sin x$
$= \frac{1}{3}u^3 + c$
$= \frac{1}{3}\sin^3 x + c$

**Exercise 3G (Page 95)**

**1 (i)** $\frac{1}{3}\sin(3x-1) + c$ **(ii)** $\dfrac{-1}{(x^2+x-1)} + c$

**(iii)** $\frac{1}{3}\tan^3 x + c$ **(iv)** $-e^{1-x} + c$

**(v)** $-\frac{1}{2}x^2\cos 2x + \frac{1}{2}x\sin 2x + \frac{1}{4}\cos 2x + c$

**(vi)** $\frac{1}{4}\sin 2x + \frac{1}{2}x + c$ **(vii)** $x\ln 2x - x + c$

**(viii)** $\dfrac{-1}{4(x^2-1)^2} + c$ **(ix)** $\frac{1}{3}(2x-3)^{\frac{3}{2}} + c$

**(x)** $\ln\left|\dfrac{x-1}{x+2}\right| - \dfrac{1}{x-1} + c$ **(xi)** $-\frac{1}{2}\cos 2x + \frac{1}{6}\cos^3 2x + c$

**(xii)** $\frac{1}{4}x^4\ln x - \frac{1}{16}x^4 + c$ **(xiii)** $\ln\left|\dfrac{x-3}{2x-1}\right| + c$

**(xiv)** $\frac{1}{2}e^{x^2+2x} + c$

**2 (i)** $\frac{8}{3}$ **(ii)** $\frac{1}{3}\ln 4$ **(iii)** $48 + 8\ln 4$

**(iv)** $\frac{5}{24}$ **(v)** $\frac{8}{3}\ln 2 - \frac{7}{9}$

**3** $\frac{4}{3}$

**4** $\frac{1}{3}(2\sqrt{2} - 1)$

**5 (i)** $\frac{\pi}{4}$ **(ii)** 0.24

**6** $\frac{1}{8}\pi - \frac{1}{4}$

**7 (i)** $-\frac{1}{2}xe^{-2x} - \frac{1}{4}e^{-2x} + c$ **(ii)** 0.112

**8 (i)** $-\frac{1}{2}\cos(2x-3)+c$ **(ii)** $\frac{3}{4}e^4+\frac{1}{4}$
**(iii)** $\frac{1}{2}\ln|x^2-9|+c$

**9 (i)** $\frac{38}{9}$ **(ii)** $\frac{1}{4}-\frac{3}{4e^2}$

**10** $\frac{1}{4}, \frac{1}{4}-\frac{3}{4e^2}$

## Chapter 4

**?** (Page 99)

At points where the rate of change of gradient is greatest.

### Exercise 4A (Page 108)

**1 (a) (i)**

| $t$ | −2 | −1.5 | −1 | −0.5 | 0 | 0.5 | 1 | 1.5 | 2 |
|---|---|---|---|---|---|---|---|---|---|
| $x$ | −4 | −3 | −2 | −1 | 0 | 1 | 2 | 3 | 4 |
| $y$ | 4 | 2.25 | 1 | 0.25 | 0 | 0.25 | 1 | 2.25 | 4 |

**(ii)**

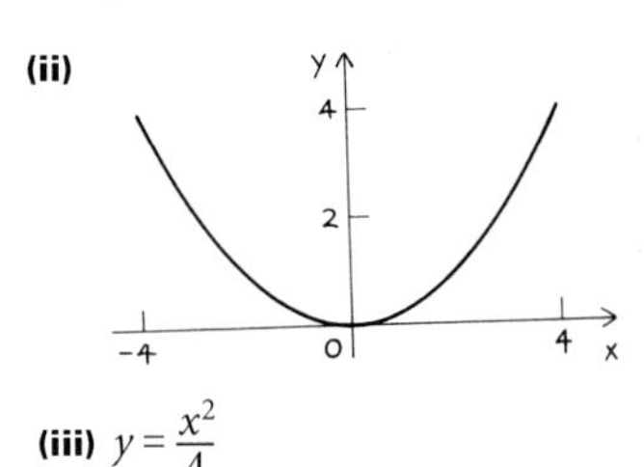


**(iii)** $y=\frac{x^2}{4}$

**(b) (i)**

| $\theta$ | 0° | 30° | 60° | 90° | 120° | 150° | 180° | 210° | 240° | 270° | 300° | 330° | 360° |
|---|---|---|---|---|---|---|---|---|---|---|---|---|---|
| $x$ | 1 | 0.5 | −0.5 | −1 | −0.5 | 0.5 | 1 | 0.5 | −0.5 | −1 | −0.5 | 0.5 | 1 |
| $y$ | 0 | 0.25 | 0.75 | 1 | 0.75 | 0.25 | 0 | 0.25 | 0.75 | 1 | 0.75 | 0.25 | 0 |

**(ii)**

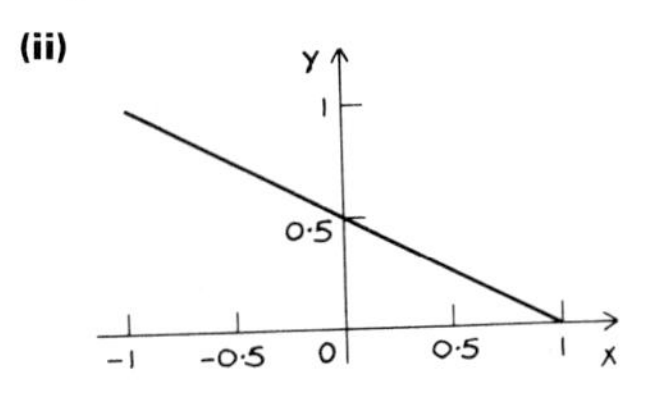


**(iii)** A segment of $y=\frac{1-x}{2}$,
where $-1 \leqslant x \leqslant 1$ and$0 \leqslant y \leqslant 1$

**(c) (i)**

| $t$ | −2 | −1.5 | −1 | −0.5 | 0 | 0.5 | 1 | 1.5 | 2 |
|---|---|---|---|---|---|---|---|---|---|
| $x$ | 4 | 2.25 | 1 | 0.25 | 0 | 0.25 | 1 | 2.25 | 4 |
| $y$ | −8 | −3.375 | −1 | −0.125 | 0 | 0.125 | 1 | 3.375 | 8 |

**(ii)**

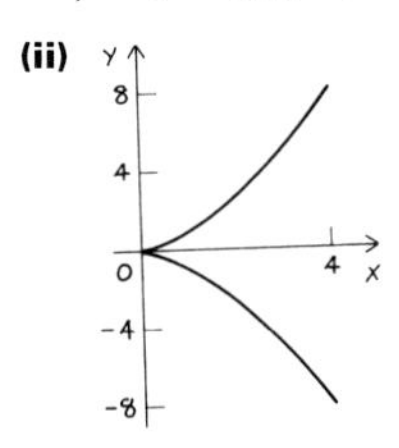


**(iii)** $y^2=x^3$

**(d) (i)**

| $\theta$ | 0° | 30° | 60° | 90° | 120° | 150° | 180° | 210° | 240° | 270° | 300° | 330° | 360° |
|---|---|---|---|---|---|---|---|---|---|---|---|---|---|
| $x$ | 0 | 0.25 | 0.75 | 1 | 0.75 | 0.25 | 0 | 0.25 | 0.75 | 1 | 0.75 | 0.25 | 0 |
| $y$ | 1 | 2 | 2.73 | 3 | 2.73 | 2 | 1 | 0 | −0.73 | −1 | −0.73 | 0 | 1 |

**(ii)**

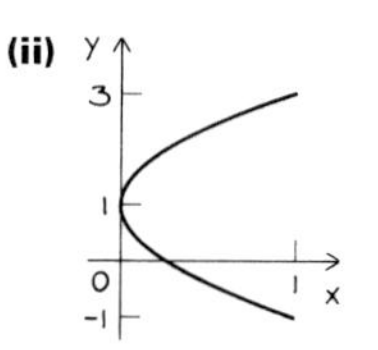


**(iii)** Part of $(y-1)^2=4x$,
where $0 \leqslant x \leqslant 1$ and $-1 \leqslant y \leqslant 3$

**(e) (i)**

| $\theta$ | 0° | 30° | 60° | 90° | 120° | 150° | 180° | 210° | 240° | 270° | 300° | 330° | 360° |
|---|---|---|---|---|---|---|---|---|---|---|---|---|---|
| $x$ | ∞ | 4 | 2.3 | 2 | 2.3 | 4 | ∞ | −4 | −2.3 | −2 | −2.3 | −4 | ∞ |
| $y$ | ∞ | 3.5 | 1.2 | 0 | −1.2 | −3.5 | ∞ | 3.5 | 1.2 | 0 | −1.2 | −3.5 | ∞ |

**(ii)**

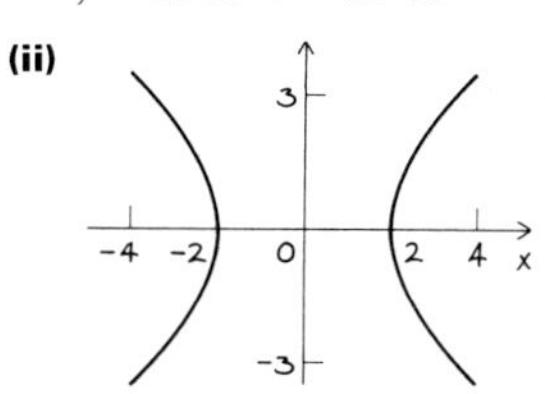


**(iii)** $x^2-y^2=4$

**(f) (i)**

| $\theta$ | 0° | 30° | 60° | 90° | 120° | 150° | 180° | 210° | 240° | 270° | 300° | 330° | 360° |
|---|---|---|---|---|---|---|---|---|---|---|---|---|---|
| $x$ | 0 | 0.5 | 1.5 | 2 | 1.5 | 0.5 | 0 | 0.5 | 1.5 | 2 | 1.5 | 0.5 | 0 |
| $y$ | 3 | 2.6 | 1.5 | 0 | −1.5 | −2.6 | −3 | −2.6 | −1.5 | 0 | 1.5 | 2.6 | 3 |

**(ii)**

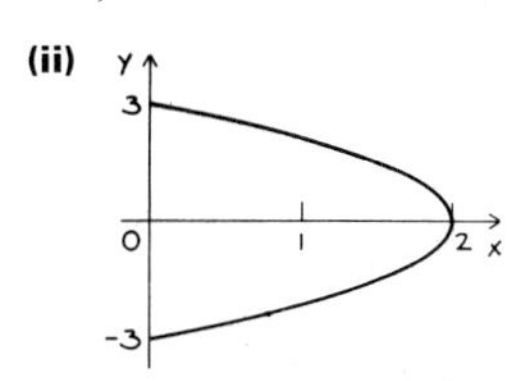


**(iii)** Part of $y^2=\frac{9}{2}(2-x)$,
where $0 \leqslant x \leqslant 2$ and $-3 \leqslant y \leqslant 3$

**(g) (i)**

| $\theta$ | 0° | 30° | 60° | 90° | 120° | 150° | 180° | 210° | 240° | 270° | 300° | 330° | 360° |
|---|---|---|---|---|---|---|---|---|---|---|---|---|---|
| $x$ | 0 | 0.6 | 1.7 | ∞ | −1.7 | −0.6 | 0 | 0.6 | 1.7 | ∞ | −1.7 | −0.6 | 0 |
| $y$ | 0 | 1.7 | −1.7 | 0 | 1.7 | −1.7 | 0 | 1.7 | −1.7 | 0 | 1.7 | −1.7 | 0 |

**(ii)**

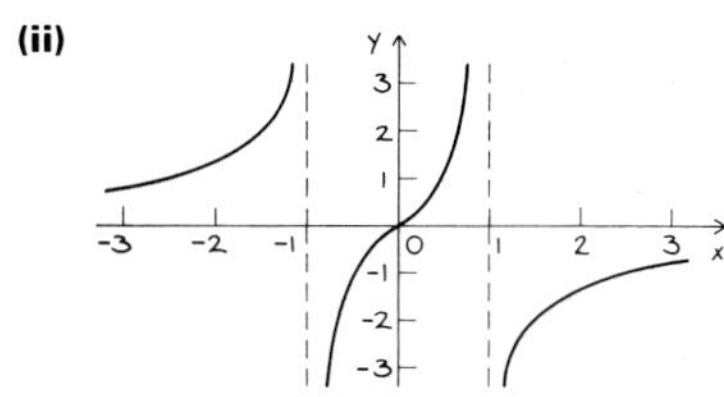


**(iii)** $y=\frac{2x}{1-x^2}$

**(h) (i)**

| $t$ | −2 | −1.5 | −1 | −0.5 | 0 | 0.5 | 1 | 1.5 | 2 |
|---|---|---|---|---|---|---|---|---|---|
| $x$ | 4 | 2.25 | 1 | 0.25 | 0 | 0.25 | 1 | 2.25 | 4 |
| $y$ | 6 | 3.75 | 2 | 0.75 | 0 | −0.25 | 0 | 0.75 | 2 |

**(ii)**

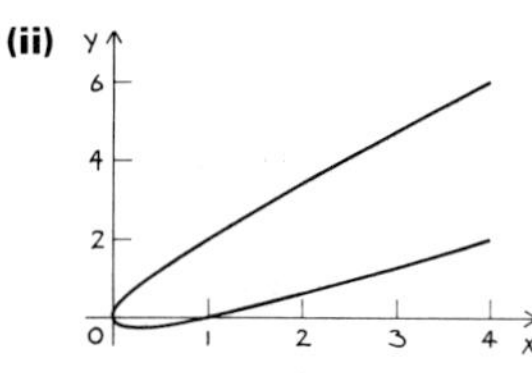


**(iii)** $y=x\pm\sqrt{x}$

**(i) (i)**

| $t$ | –2 | –1.5 | –1 | –0.5 | 0 | 0.5 | 1 | 1.5 | 2 |
|---|---|---|---|---|---|---|---|---|---|
| $x$ | 2 | 3 | ∞ | –1 | 0 | 0.33 | 0.5 | 0.6 | 0.7 |
| $y$ | –0.7 | –0.6 | –0.5 | –0.33 | 0 | 1 | ∞ | –3 | –2 |

**(ii)**

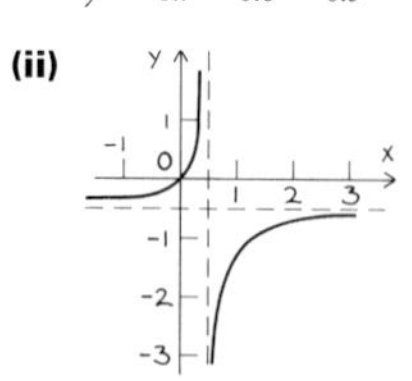


**(iii)** $y = \dfrac{x}{1-2x}$

**2 (i)** $x^2 + y^2 = 25$

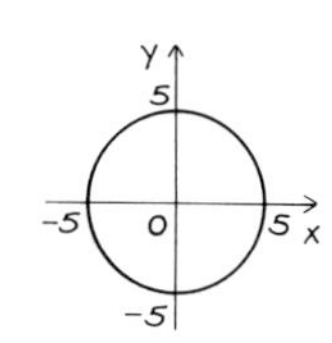


**(ii)** $\dfrac{x^2}{9} + \dfrac{y^2}{4} = 1$

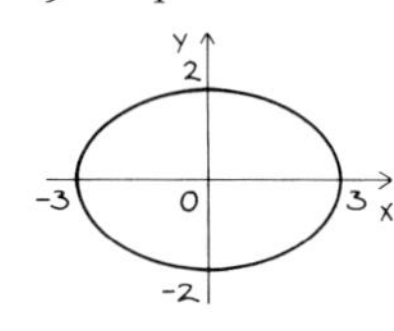


**(iii)** Segment of $y = x - 3$, where $1 \leqslant x \leqslant 7$ and $-2 \leqslant y \leqslant 4$

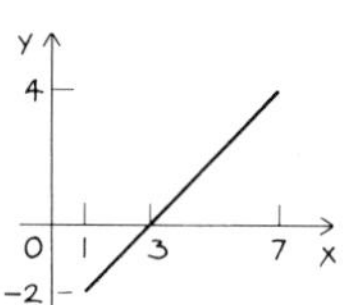


**(iv)** $(x+1)^2 + (y-3)^2 = 4$

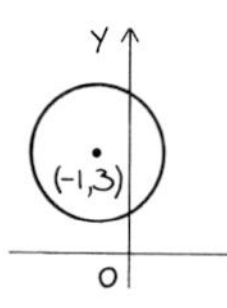


**3 (i)** **(a)** $xy = 1$ **(b)** $xy = 16$

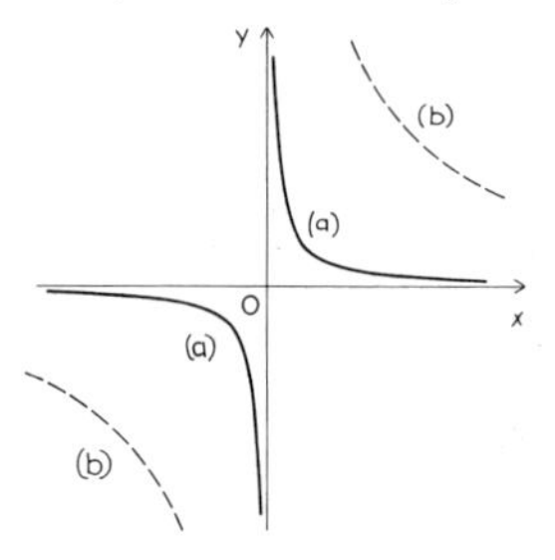


**4 (i)** $\dfrac{x^2}{25} + \dfrac{y^2}{9} = 1$

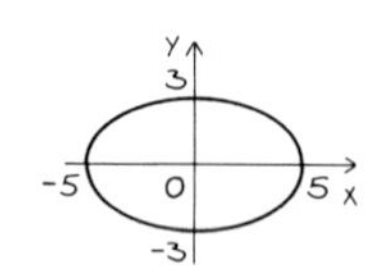


**(ii)** Inscribed circle $x = 3\cos\theta$, $y = 3\sin\theta$; circumscribing circle $x = 5\cos\theta$, $y = 5\sin\theta$

**5 (i)**

| $t$ | –2 | –1.5 | –1 | –0.5 | 0 | 0.5 | 1 | 1.5 | 2 |
|---|---|---|---|---|---|---|---|---|---|
| $x$ | 4 | 2.25 | 1 | 0.25 | 0 | 0.25 | 1 | 2.25 | 4 |
| $y$ | 16 | 5.0625 | 1 | 0.0625 | 0 | 0.0625 | 1 | 5.0625 | 16 |

**(ii)**

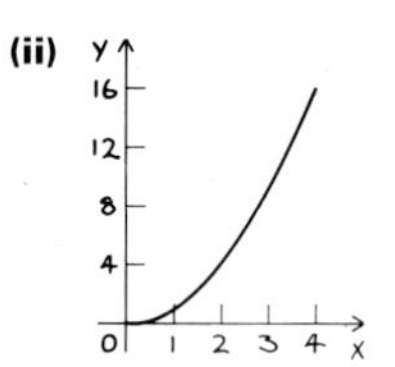


**(iii)** Because it should also state 'for $x \geqslant 0$'

**6 (i)** $y = \dfrac{x}{2} - \dfrac{x^2}{80}$

**(ii)**

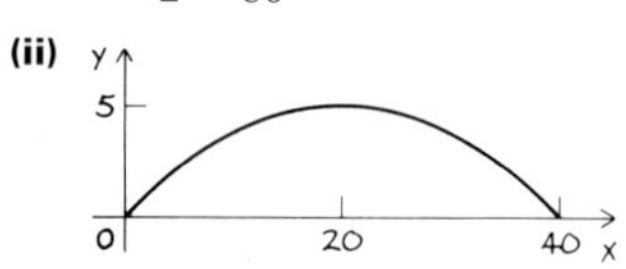


**7 (i)**

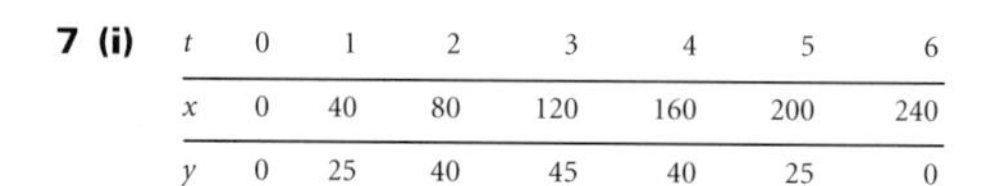

| $t$ | 0 | 1 | 2 | 3 | 4 | 5 | 6 |
|---|---|---|---|---|---|---|---|
| $x$ | 0 | 40 | 80 | 120 | 160 | 200 | 240 |
| $y$ | 0 | 25 | 40 | 45 | 40 | 25 | 0 |

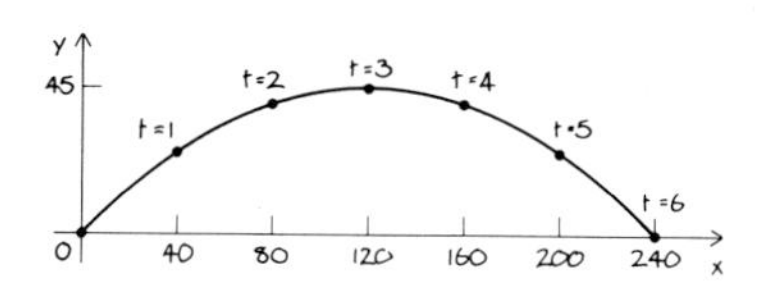


**(ii)** 240m

**(iii)**

| $t$ | 0 | 1 | 2 | 3 | 4 | 5 | 6 |
|---|---|---|---|---|---|---|---|
| $x$ | 0 | 39 | 76 | 111 | 144 | 175 | 204 |
| $y$ | 0 | 25 | 40 | 45 | 40 | 25 | 0 |

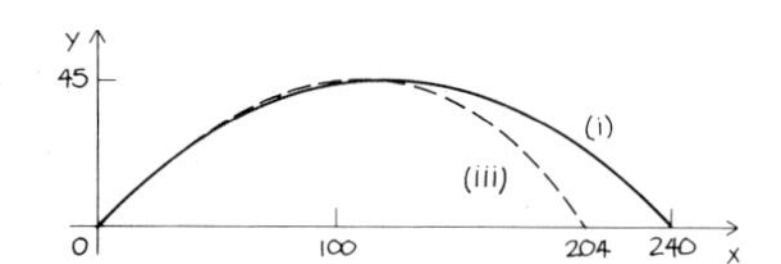


**(iv)** 36m

**8 (i)**

| $t$ | –4 | –3 | –2 | –1 | 0 | 1 | 2 | 3 | 4 |
|---|---|---|---|---|---|---|---|---|---|
| $x$ | 9 | 4 | 1 | 0 | 1 | 4 | 9 | 16 | 25 |
| $y$ | –5 | –4 | –3 | –2 | –1 | 0 | 1 | 2 | 3 |

**(ii)**

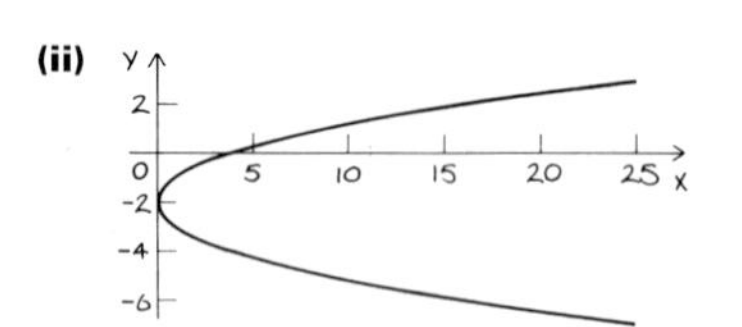


**(iii)** $y = -2$ **(iv)** $x = (y+2)^2$

**9 (i)**

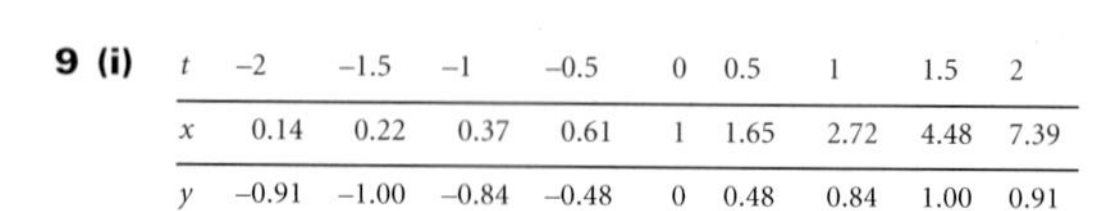

| $t$ | –2 | –1.5 | –1 | –0.5 | 0 | 0.5 | 1 | 1.5 | 2 |
|---|---|---|---|---|---|---|---|---|---|
| $x$ | 0.14 | 0.22 | 0.37 | 0.61 | 1 | 1.65 | 2.72 | 4.48 | 7.39 |
| $y$ | –0.91 | –1.00 | –0.84 | –0.48 | 0 | 0.48 | 0.84 | 1.00 | 0.91 |

**(ii)** $x > 0$

**(iii)**

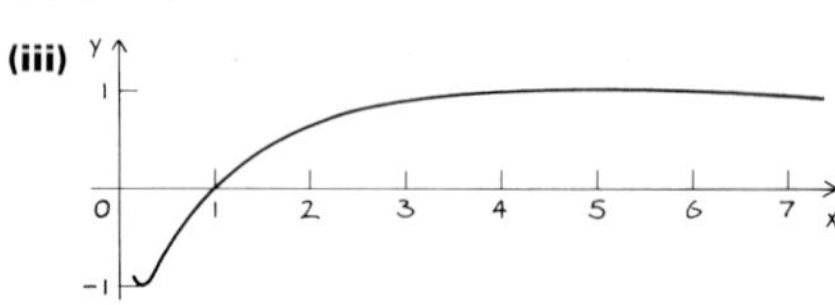

**(iv)** The graph oscillates infinitely many times from –1 to +1 for $t < -2$, i.e. where $0 < x < 0.14$. For $t > 2$ the graph oscillates infinitely many times from –1 to +1, but successive distances between a maximum and a minimum become increasingly large.

**10 (i)**

| $\theta$ | 0 | $\frac{\pi}{3}$ | $\frac{2\pi}{3}$ | $\pi$ | $\frac{4\pi}{3}$ | $\frac{5\pi}{3}$ | $2\pi$ | $\frac{7\pi}{3}$ | $\frac{8\pi}{3}$ | $3\pi$ | $\frac{10\pi}{3}$ | $\frac{11\pi}{3}$ | $4\pi$ |
|---|---|---|---|---|---|---|---|---|---|---|---|---|---|
| $x$ | 0 | $0.2a$ | $1.2a$ | $3.1a$ | $5.1a$ | $6.1a$ | $6.3a$ | $6.5a$ | $7.5a$ | $9.4a$ | $11.3a$ | $12.4a$ | $12.6a$ |
| $y$ | 0 | $0.5a$ | $1.5a$ | $2a$ | $1.5a$ | $0.5a$ | 0 | $0.5a$ | $1.5a$ | $2a$ | $1.5a$ | $0.5a$ | 0 |

| $\theta$ | $\frac{13\pi}{3}$ | $\frac{14\pi}{3}$ | $5\pi$ | $\frac{16\pi}{3}$ | $\frac{17\pi}{3}$ | $6\pi$ |
|---|---|---|---|---|---|---|
| $x$ | $12.7a$ | $13.8a$ | $15.7a$ | $17.6a$ | $18.7a$ | $18.8a$ |
| $y$ | $0.5a$ | $1.5a$ | $2a$ | $1.5a$ | $0.5a$ | 0 |

**(ii)**

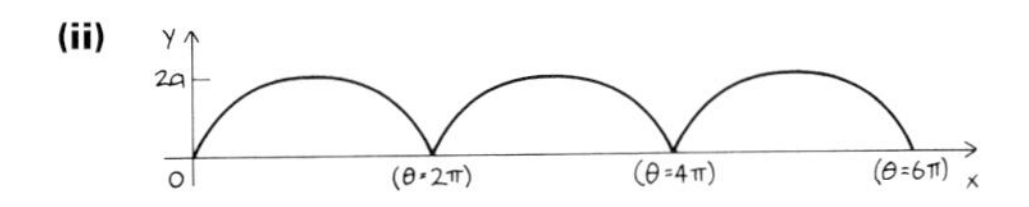


**(iii)** Periodic

**11 (i)**

| $\theta$ | 0 | $\frac{\pi}{3}$ | $\frac{\pi}{2}$ | $\frac{2\pi}{3}$ | $\pi$ | $\frac{4\pi}{3}$ | $\frac{3\pi}{2}$ | $\frac{5\pi}{3}$ | $2\pi$ |
|---|---|---|---|---|---|---|---|---|---|
| $x$ | $a$ | $0.13a$ | 0 | $-0.13a$ | $-a$ | $-0.13a$ | 0 | $0.13a$ | $a$ |
| $y$ | 0 | $0.65a$ | $a$ | $0.65a$ | 0 | $-0.65a$ | $-a$ | $-0.65a$ | 0 |

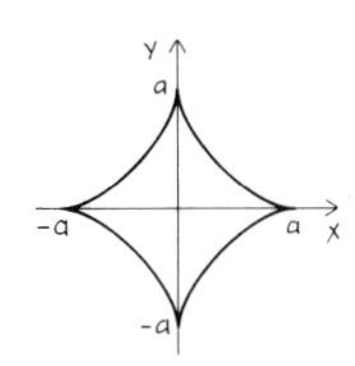


**(iii) (a)** The larger the value of $n$, the closer the curve is to the axes. If the power is even, the curve is only in the first quadrant.

**(b)** If the power is odd, the curve is in all four quadrants.

## Exercise 4B (Page 116)

**1 (i)** $t$ **(ii)** $\dfrac{1 + \cos\theta}{1 + \sin\theta}$ **(iii)** $\dfrac{t^2 + 1}{t^2 - 1}$ **(iv)** $-\frac{2}{3}\cot\theta$

**(v)** $\dfrac{t - 1}{t + 1}$ **(vi)** $-\tan\theta$ **(vii)** $\dfrac{1}{2e^t}$ **(viii)** $\dfrac{(1 + t)^2}{(1 - t)^2}$

**2 (i)** 6 **(ii)** $y = 6x - \sqrt{3}$ **(iii)** $3x + 18y - 19\sqrt{3} = 0$

**3 (i)** $(\frac{1}{4}, 0)$ **(ii)** 2 **(iii)** $y = 2x - \frac{1}{2}$ **(iv)** $(0, -\frac{1}{2})$

**4 (i)** $x - ty + at^2 = 0$ **(ii)** $tx + y = at^3 + 2at$
**(iii)** $(at^2 + 2a, 0)$, $(0, at^3 + 2at)$

**6 (i)** $-\dfrac{b}{at^2}$ **(ii)** $at^2y + bx = 2abt$

**(iii)** $X(2at, 0)$, $Y\left(0, \dfrac{2b}{t}\right)$ **(iv)** Area $= 2ab$

**7 (i)**

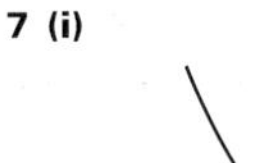

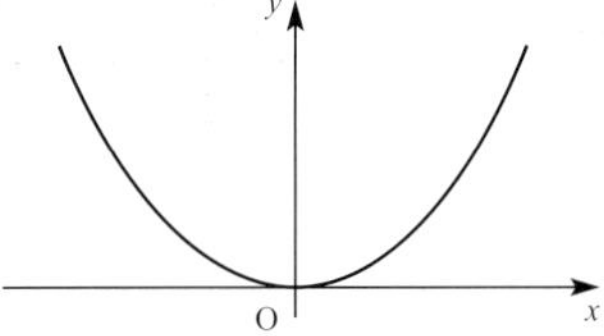


**(iii)** $y = tx - 2t^2$
**(iv)** $[2(t_1 + t_2), 2t_1t_2]$
**(v)** $x = 4$

**8 (i)** $t = 1$
**(iii)** $x + y = 3$
**(v)** $(-8, -5)$

**9 (i)** $t = -2$
**(iii)** $y = 2x - 6$
**(iv)** $(-5, 9)$

**10 (i)** $-\dfrac{3\cos t}{4\sin t}$ **(ii)** $3x\cos t + 4y\sin t = 12$
**(iii)** $t = 0.6435 + n\pi$

**11 (i)** $x\cos\theta + y\sin\theta = 3\sin\theta + 3\cos\theta + 2$
**(iii)** 2.85, 5.01 radians
**(iv)**

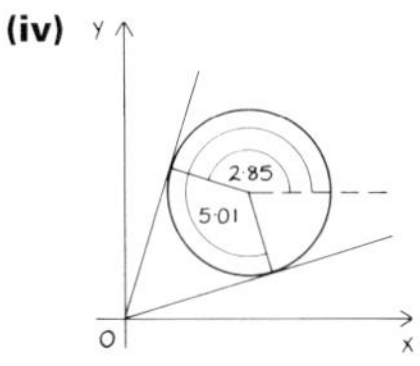


**12 (i)** $-\dfrac{\cos\theta}{\sin\theta}$ **(ii)** $y\cos\theta - x\sin\theta = 5\cos\theta - 2\sin\theta$

**13 (i)** $\dfrac{x^2}{9} + \dfrac{y^2}{4} = 1$

**(ii)**

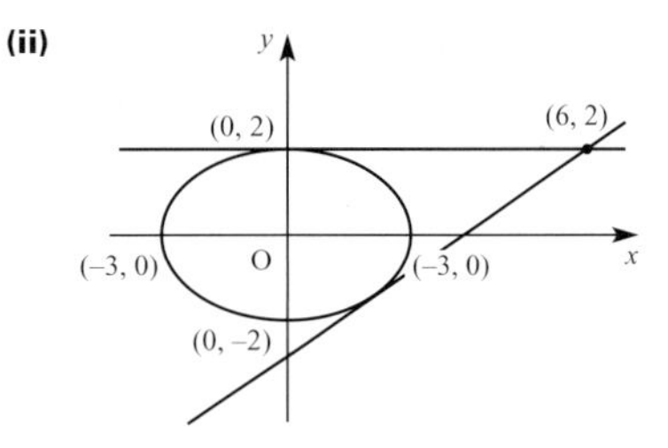


**(iii)** $-\dfrac{2\cos\theta}{3\sin\theta}$

**(v)** $\theta = 1.57$ or $5.64$ (2d.p.)

**14 (i) (a)** $\dfrac{x^2}{16} + \dfrac{y^2}{9} = 1$ **(b)** $20\sin(\theta + 0.9273)$
**(c)** max. $L = 20$ when $\theta = 0.6435$

**(ii) (a)** $-\dfrac{2}{\cos\theta}$ **(b)** $-\dfrac{3\cos\theta}{4\sin\theta}$
**(c)** $\theta = 0.34$

**15 (i)** $t = \dfrac{1-x}{x}$

**(ii)** $\dfrac{dy}{dt} = \dfrac{2t}{(1+t)^2(1-t)^2}$; $\dfrac{dx}{dt} = -\dfrac{1}{(1+t)^2}$

**(iii)** $y = \dfrac{1}{1-t}$

**16 (i)**

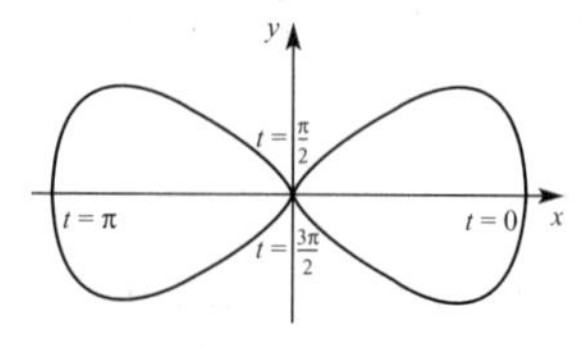


**(ii)** $\dfrac{dy}{dx} = -\dfrac{\cos 2t}{\sin t}$

**(iv)** $t = 0.253$ or $2.889$ (3d.p.)

**17 (i)** $x = \sqrt{5}\cos(\theta - 0.4636)$; $-\sqrt{5} \leqslant x \leqslant \sqrt{5}$

**(ii)** $\frac{1}{2}$ **(iv)** 3; 1

## Chapter 5

**?** (Page 123)

To find the distance between the two vapour trails you need two pieces of information for each of them: either two points that it goes through, or else one point and its direction. All of these need to be in three dimensions. However, if you want to find the closest approach of the two aircraft you also need to know, for each of them, the time at which it was at a given point on its trail and the speed at which it was travelling. (This answer assumes constant speeds and directions.)

### Exercise 5A (Page 129)

**1 (i)** $3\mathbf{i} + 2\mathbf{j}$ **(ii)** $5\mathbf{i} - 4\mathbf{j}$ **(iii)** $3\mathbf{i}$
**(iv)** $-3\mathbf{i} - \mathbf{j}$ **(v)** $2\mathbf{j}$

**2 (i)**

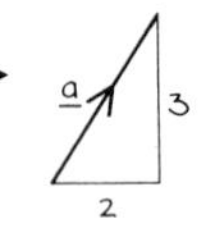


$(\sqrt{13}, 56.3°)$

**(ii)**

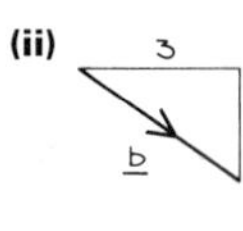


$(\sqrt{13}, -33.7°)$

**(iii)**

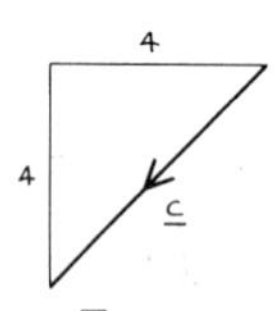


$(4\sqrt{2}, -135°)$

**(iv)** $(\sqrt{5}, 116.6°)$

**(v)**

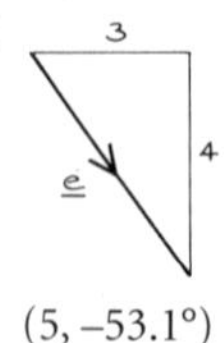


$(5, -53.1°)$

**3 (i)**

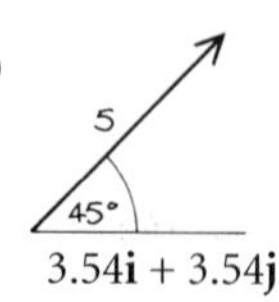


$3.54\mathbf{i} + 3.54\mathbf{j}$

**(ii)**

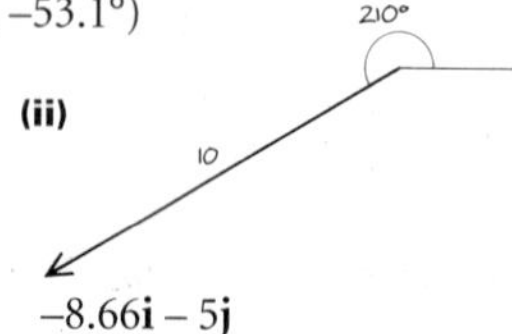


$-8.66\mathbf{i} - 5\mathbf{j}$

**(iii)** $4\mathbf{j}$

**(iv)**

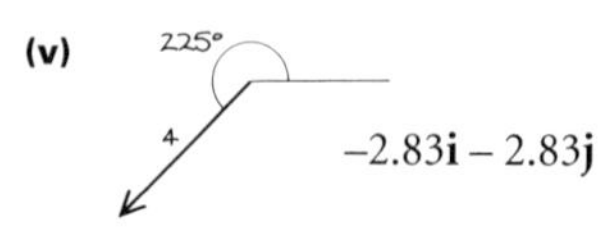


**(v)** $-2.83\mathbf{i} - 2.83\mathbf{j}$

**4 (i)** $2\mathbf{i} - 2\mathbf{j}$ **(ii)** $2\mathbf{i}$ **(iii)** $-4\mathbf{j}$ **(iv)** $4\mathbf{j}$
**(v)** $-\mathbf{i} + \mathbf{j}$ **(vi)** $\mathbf{i} - \mathbf{j}$ **(vii)** $8\mathbf{i}$ **(viii)** $-8\mathbf{i}$
**(ix)** $2\mathbf{i} - 4\mathbf{j}$ **(x)** $4\mathbf{i} + 4\mathbf{j}$

**5 (i)** A: $2\mathbf{i} + 3\mathbf{j}$, C: $-2\mathbf{i} + \mathbf{j}$ **(ii)** $\overrightarrow{AB} = -2\mathbf{i} + \mathbf{j}$, $\overrightarrow{CB} = 2\mathbf{i} + 3\mathbf{j}$
**(iii) (a)** $\overrightarrow{AB} = \overrightarrow{OC}$ **(b)** $\overrightarrow{CB} = \overrightarrow{OA}$
**(iv)** A parallelogram

### Exercise 5B (Page 135)

**1 (i)** $\begin{pmatrix}6\\8\end{pmatrix}$ **(ii)** $\begin{pmatrix}1\\1\end{pmatrix}$ **(iii)** $\begin{pmatrix}0\\0\end{pmatrix}$ **(iv)** $\begin{pmatrix}8\\-1\end{pmatrix}$ **(v)** $-3\mathbf{j}$

**2 (i)** $2\mathbf{i} + 3\mathbf{j}$ **(ii)** $\mathbf{i}$ **(iii)** $\mathbf{j}$ **(iv)** $3\mathbf{i} + 2\mathbf{j}$ **(v)** $\mathbf{0}$

**3 (i) (a)** $\mathbf{b}$ **(b)** $\mathbf{a} + \mathbf{b}$ **(c)** $-\mathbf{a} + \mathbf{b}$
**(ii) (a)** $\frac{1}{2}(\mathbf{a} + \mathbf{b})$ **(b)** $\frac{1}{2}(-\mathbf{a} + \mathbf{b})$
**(iii)** PQRS is any parallelogram and $\overrightarrow{PM} = \frac{1}{2}\overrightarrow{PR}$, $\overrightarrow{QM} = \frac{1}{2}\overrightarrow{QS}$

**4 (i) (a)** $\mathbf{i}$ **(b)** $2\mathbf{i}$ **(c)** $\mathbf{i} - \mathbf{j}$ **(d)** $-\mathbf{i} - 2\mathbf{j}$
**(ii)** $|\overrightarrow{AB}| = |\overrightarrow{BC}| = \sqrt{2}$, $|\overrightarrow{AD}| = |\overrightarrow{CD}| = \sqrt{5}$

**5 (i)** $-\mathbf{p} + \mathbf{q}$, $\frac{1}{2}\mathbf{p} - \frac{1}{2}\mathbf{q}$, $-\frac{1}{2}\mathbf{p}$, $-\frac{1}{2}\mathbf{q}$
**(ii)** $\overrightarrow{NM} = \frac{1}{2}\overrightarrow{BC}$, $\overrightarrow{NL} = \frac{1}{2}\overrightarrow{AC}$, $\overrightarrow{ML} = \frac{1}{2}\overrightarrow{AB}$

**6 (i)** $\begin{pmatrix}\frac{2}{\sqrt{13}}\\ \frac{3}{\sqrt{13}}\end{pmatrix}$ **(ii)** $\frac{3}{5}\mathbf{i} + \frac{4}{5}\mathbf{j}$ **(iii)** $\begin{pmatrix}\frac{-1}{\sqrt{2}}\\ \frac{-1}{\sqrt{2}}\end{pmatrix}$

**(iv)** $\frac{5}{13}\mathbf{i} - \frac{12}{13}\mathbf{j}$ **(v)** $\mathbf{i}$ **(vi)** $\begin{pmatrix}\frac{-1}{\sqrt{5}}\\ \frac{2}{\sqrt{5}}\end{pmatrix}$

**(vii)** $\begin{pmatrix}\frac{-1}{\sqrt{5}}\\ \frac{2}{\sqrt{5}}\end{pmatrix}$ **(viii)** $\begin{pmatrix}\frac{1}{\sqrt{5}}\\ \frac{2}{\sqrt{5}}\end{pmatrix}$

**(ix)** $\begin{pmatrix}\cos\alpha\\ \sin\alpha\end{pmatrix}$ **(x)** $\begin{pmatrix}\cos\beta\\ \sin\beta\end{pmatrix}$

### Activity (Page 139)

**(ii)** $\begin{pmatrix}-2\\-9\end{pmatrix}$, $\begin{pmatrix}0\\-5\end{pmatrix}$, $\begin{pmatrix}2\\-1\end{pmatrix}$, $\begin{pmatrix}3\\1\end{pmatrix}$, $\begin{pmatrix}3\frac{1}{2}\\2\end{pmatrix}$, $\begin{pmatrix}4\\3\end{pmatrix}$, $\begin{pmatrix}8\\11\end{pmatrix}$
**(iv)** $0, 1, \frac{1}{2}, \frac{3}{4}$
**(v) (a)** It lies between A and B.
**(b)** It lies beyond B.
**(c)** It lies beyond A.

### Activity (Page 142)

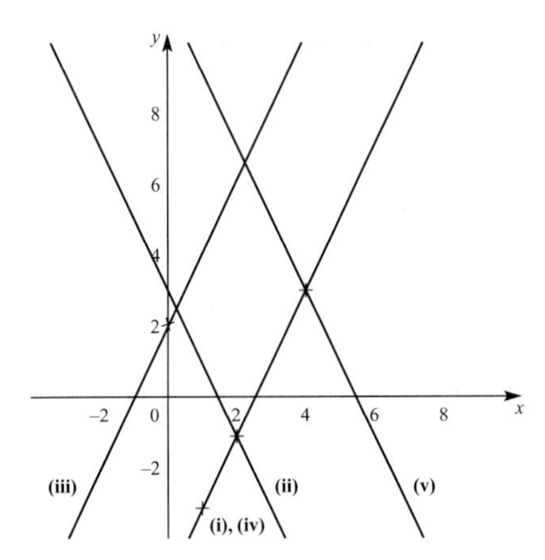


**(i)** and **(iv)** are the same since **(a)** putting $\lambda = -1$ in **(i)** gives $\begin{pmatrix}1\\-3\end{pmatrix}$ **(b)** $\begin{pmatrix}1\\2\end{pmatrix}$ is parallel to $\begin{pmatrix}3\\6\end{pmatrix}$.

**(iii)** is parallel to **(i)** since the direction vector is the same.

**(iv)** is parallel to **(ii)** since $\begin{pmatrix}-1\\2\end{pmatrix} = -\begin{pmatrix}1\\-2\end{pmatrix}$.

### Exercise 5C (Page 145)

**1 (a) (i)** $2\mathbf{i} + 8\mathbf{j}$ **(ii)** 68 **(iii)** $3\mathbf{i} + 7\mathbf{j}$

**(b) (i)** $-4\mathbf{i} - 3\mathbf{j}$ **(ii)** 5√ **(iii)** $2\mathbf{i} + 1.5\mathbf{j}$

**(c) (i)** $6\mathbf{i} + 8\mathbf{j}$ **(ii)** 10 **(iii)** $\mathbf{i} + 3\mathbf{j}$

**(d) (i)** $6\mathbf{i} - 8\mathbf{j}$ **(ii)** 10 **(iii)** $\mathbf{0}$

**(e) (i)** $5\mathbf{i} + 12\mathbf{j}$ **(ii)** 13 **(iii)** $-7.5\mathbf{i} - 2\mathbf{j}$

**2** *Note: These answers are not unique.*

**(i)** $\mathbf{r} = \begin{pmatrix}2\\1\end{pmatrix} + \lambda\begin{pmatrix}1\\2\end{pmatrix}$ **(ii)** $\mathbf{r} = \begin{pmatrix}3\\5\end{pmatrix} + \lambda\begin{pmatrix}-1\\1\end{pmatrix}$ **(iii)** $\mathbf{r} = \begin{pmatrix}-6\\-6\end{pmatrix} + \lambda\begin{pmatrix}1\\1\end{pmatrix}$

**(iv)** $\mathbf{r} = \begin{pmatrix}5\\3\end{pmatrix} + \lambda\begin{pmatrix}1\\1\end{pmatrix}$ **(v)** $\mathbf{r} = \lambda\begin{pmatrix}2\\1\end{pmatrix}$ **(vi)** $\mathbf{r} = \lambda\begin{pmatrix}-1\\4\end{pmatrix}$

**(vii)** $\mathbf{r} = \lambda\begin{pmatrix}-1\\4\end{pmatrix}$ **(viii)** $\mathbf{r} = \begin{pmatrix}3\\-12\end{pmatrix} + \lambda\begin{pmatrix}-1\\4\end{pmatrix}$

**3 (i)** $y = 3x - 1$ **(ii)** $y = \frac{1}{2}x + 1$ **(iii)** $y = x - 1$

**(iv)** $y = x - 1$ **(v)** $y = 5$ ($x$ may take any value)

**4** *Note: These answers are not unique.*

**(i)** $\mathbf{r} = \begin{pmatrix}0\\3\end{pmatrix} + \lambda\begin{pmatrix}1\\2\end{pmatrix}$ **(ii)** $\mathbf{r} = \begin{pmatrix}0\\-4\end{pmatrix} + \lambda\begin{pmatrix}1\\1\end{pmatrix}$

**(iii)** $\mathbf{r} = \begin{pmatrix}0\\-1\end{pmatrix} + \lambda\begin{pmatrix}2\\1\end{pmatrix}$ **(iv)** $\mathbf{r} = \lambda\begin{pmatrix}-4\\1\end{pmatrix}$

**(v)** $\mathbf{r} = \begin{pmatrix}0\\4\end{pmatrix} + \lambda\begin{pmatrix}-2\\1\end{pmatrix}$

**5 (i)** $\begin{pmatrix}4\\1\end{pmatrix}$ **(ii)** $\begin{pmatrix}5\\5\end{pmatrix}$ **(iii)** $\begin{pmatrix}12\\17\end{pmatrix}$ **(iv)** $\begin{pmatrix}-5\\6\end{pmatrix}$ **(v)** $\begin{pmatrix}6\\3\end{pmatrix}$

**6 (i)** 12.8 km **(ii)** $20 \text{ km h}^{-1}$, $5 \text{ km h}^{-1}$

**(iii)** After 40 minutes there is a collision.

**7 (i)** $\overrightarrow{OL} = \begin{pmatrix}10\\4.5\end{pmatrix}$; $\overrightarrow{OM} = \begin{pmatrix}7\\3.5\end{pmatrix}$; $\overrightarrow{ON} = \begin{pmatrix}4\\1\end{pmatrix}$

**(ii)** AL: $\mathbf{r} = \begin{pmatrix}1\\0\end{pmatrix} + \lambda\begin{pmatrix}2\\1\end{pmatrix}$; BM: $\mathbf{r} = \begin{pmatrix}7\\2\end{pmatrix} + \mu\begin{pmatrix}0\\1\end{pmatrix}$;
CN: $\mathbf{r} = \begin{pmatrix}13\\7\end{pmatrix} + \nu\begin{pmatrix}3\\2\end{pmatrix}$

**(iii) (a)** (7, 3) **(b)** (7, 3)

**(iii)** The lines AL, BM and CN are concurrent. (They are the medians of the triangle, and this result holds for the medians of any triangle.)

### Exercise 5D (Page 150)

**1 (i)** 42.3° **(ii)** 90° **(iii)** 18.4°

**(iv)** 31.0° **(v)** 90° **(vi)** 180°

**2 (i)** $\mathbf{r} = \begin{pmatrix}1\\0\end{pmatrix} + \lambda\begin{pmatrix}2\\1\end{pmatrix}$ **(ii)** $\mathbf{r} = \begin{pmatrix}6\\1\end{pmatrix} + \mu\begin{pmatrix}1\\2\end{pmatrix}$

**(iii)** $\begin{pmatrix}7\\3\end{pmatrix}$ **(iv)** 36.9°

**3 (i)** Parallelogram: AB||CD, BC||DA

**(ii)** A(5, 2); B(1, 1); C(2, 4); D(6, 5)

**(iii)** 57.5°, 122.5°

**4 (i)** $\begin{pmatrix}3\\1\end{pmatrix}, \begin{pmatrix}-1\\3\end{pmatrix}$ **(ii)** $\overrightarrow{BA} \cdot \overrightarrow{BC} = 0$

**(iii)** $|\overrightarrow{AB}| = |\overrightarrow{BC}| = \sqrt{10}$ **(iv)** (2, 5)

**5 (i)** $\overrightarrow{PQ} = -4\mathbf{i} + 2\mathbf{j}$; $\overrightarrow{RQ} = 4\mathbf{i} + 8\mathbf{j}$

**(ii)** 26.6° **(iii)** $3\mathbf{i} + 7\mathbf{j}$ **(iv)** 53.1°

### Exercise 5E (Page 159)

*Note: Many of these answers are not unique.*

**1 (i)** $\mathbf{r} = \begin{pmatrix}2\\4\\-1\end{pmatrix} + \lambda\begin{pmatrix}3\\6\\4\end{pmatrix}$ **(ii)** $\mathbf{r} = \begin{pmatrix}1\\0\\-1\end{pmatrix} + \lambda\begin{pmatrix}1\\0\\0\end{pmatrix}$

**(iii)** $\mathbf{r} = \begin{pmatrix}1\\0\\4\end{pmatrix} + \lambda\begin{pmatrix}5\\3\\-6\end{pmatrix}$ **(iv)** $\mathbf{r} = \begin{pmatrix}0\\0\\1\end{pmatrix} + \lambda\begin{pmatrix}2\\1\\3\end{pmatrix}$

**(v)** $\mathbf{r} = \lambda\begin{pmatrix}1\\2\\3\end{pmatrix}$

**2 (i)** $\dfrac{x-2}{3} = \dfrac{y-4}{6} = \dfrac{z+1}{4}$ **(ii)** $x - 1 = \dfrac{y}{3} = \dfrac{z+1}{4}$

**(iii)** $x - 3 = \dfrac{z-4}{2}$ and $y = 0$

**(iv)** $\dfrac{x}{2} = \dfrac{z+1}{4}$ and $y = 4$ **(v)** $x = -2$ and $z = 3$

**3 (i)** $\mathbf{r} = \begin{pmatrix}3\\-2\\1\end{pmatrix} + \lambda\begin{pmatrix}5\\3\\4\end{pmatrix}$ **(ii)** $\mathbf{r} = \begin{pmatrix}-6\\0\\-4\end{pmatrix} + \lambda\begin{pmatrix}6\\2\\3\end{pmatrix}$

**(iii)** $\mathbf{r} = \begin{pmatrix}0\\0\\-1\end{pmatrix} + \lambda\begin{pmatrix}1\\2\\3\end{pmatrix}$ **(iv)** $\mathbf{r} = \lambda\begin{pmatrix}1\\1\\1\end{pmatrix}$

**(v)** $\mathbf{r} = \begin{pmatrix}2\\0\\0\end{pmatrix} + \lambda\begin{pmatrix}0\\1\\1\end{pmatrix}$

**4 (i)** 29.0° **(ii)** 76.2° **(iii)** 162.0°

**5 (i)** 53.6° **(ii)** 81.8° **(iii)** 8.7°

**6 (i)** (0, 4, 3) **(ii)** $\begin{pmatrix}-5\\4\\3\end{pmatrix}$, $\sqrt{50}$

**(iii)** $\mathbf{r} = \begin{pmatrix}5\\0\\0\end{pmatrix} + \lambda\begin{pmatrix}-5\\4\\3\end{pmatrix}$ **(iv)** $\begin{pmatrix}3\frac{3}{4}\\1\\\frac{3}{4}\end{pmatrix}$, 63.4°

**(v)** Spider is then at P(2.5, 2, 1.5) and $\overrightarrow{OP}.\overrightarrow{AG} = 0$, $|\overrightarrow{OP}| = 3.54$

**7 (i)** A(4, 0, 0), F(4, 0, 3) **(ii)** 114.1°, 109.5°
**(iii)** They touch but are not perpendicular

**8 (i)** $\begin{pmatrix}-0.25\\0\\0\end{pmatrix}$ **(ii)** (0, 0.05, 1.1)

**(iii)** DE: $\mathbf{r} = \begin{pmatrix}0\\0\\1\end{pmatrix} + \lambda\begin{pmatrix}1\\0\\0\end{pmatrix}$

EF: $\mathbf{r} = \begin{pmatrix}0.25\\0\\1\end{pmatrix} + \lambda\begin{pmatrix}0\\1\\2\end{pmatrix}$

**?** **(Page 162)**

A three-legged stool is the more stable. Three points, such as the ends of the legs, define a plane but a fourth will not, in general, be in the same plane. So the ends of the legs of a three-legged stool lie in a plane but those of a four-legged stool do not. The four-legged stool will rest on three legs but can rock on to a different three.

**?** **(Page 164)**

**(i)** 90° with all lines.
**(ii)** No, so long as the pencil remains perpendicular to the table.

**Exercise 5F (Page 169)**

**1 (i)** $\overrightarrow{AB} = \begin{pmatrix}-2\\-2\\-6\end{pmatrix}$; $\overrightarrow{AC} = \begin{pmatrix}1\\-2\\-1\end{pmatrix}$

**(ii)** $\mathbf{r} = \begin{pmatrix}0\\1\\1\end{pmatrix} + \lambda\begin{pmatrix}-2\\-2\\-6\end{pmatrix} + \mu\begin{pmatrix}1\\-2\\-1\end{pmatrix}$

**(iv)** The vector $\begin{pmatrix}5\\4\\-3\end{pmatrix}$ is perpendicular to the plane ABC

**2 (i)** $\overrightarrow{LM} = \begin{pmatrix}2\\2\\-2\end{pmatrix}$; $\overrightarrow{LN} = \begin{pmatrix}5\\2\\-1\end{pmatrix}$ **(iii)** $x - 4y - 3z = -2$

**3 (iii)** The plane is parallel to the *yz* plane and passes through (3, 0, 0)

**4 (iii)** B

**5 (iii)** Three points define a plane **(iv)** (1, 0, –1)

**6 (i)** (0, 1, 3) **(ii)** (1, 1, 1) **(iii)** (8, 4, 2)
**(iv)** (0, 0, 0) **(v)** (11, 19, –10)

**7 (i) (a)** $\mathbf{r} = \begin{pmatrix}2\\2\\3\end{pmatrix} + \lambda\begin{pmatrix}1\\-1\\2\end{pmatrix}$ **(b)** (1, 3, 1) **(c)** $\sqrt{6}$

**(ii) (a)** $\mathbf{r} = \begin{pmatrix}2\\3\\0\end{pmatrix} + \lambda\begin{pmatrix}2\\5\\3\end{pmatrix}$ **(b)** (1, 0.5, –1.5) **(c)** 3.08

**(iii) (a)** $\mathbf{r} = \begin{pmatrix}3\\1\\3\end{pmatrix} + \lambda\begin{pmatrix}1\\0\\0\end{pmatrix}$ **(b)** (0, 1, 3) **(c)** 3

**(iv) (a)** $\mathbf{r} = \begin{pmatrix}2\\1\\0\end{pmatrix} + \lambda\begin{pmatrix}3\\-4\\1\end{pmatrix}$

**(b)** (2, 1, 0): A is on the plane **(c)** 0

**(v) (a)** $\mathbf{r} = \lambda\begin{pmatrix}1\\1\\1\end{pmatrix}$ **(b)** (2, 2, 2) **(c)** $\sqrt{12}$

**8 (i)** $x + 2y + 3z = 25$ **(ii)** 206 = 150 + 56
**(iii)** W is on the plane; $\overrightarrow{UW}.\overrightarrow{UV} = 0$

**9 (i)** $\mathbf{r} = \begin{pmatrix}13\\5\\0\end{pmatrix} + \lambda\begin{pmatrix}3\\1\\-2\end{pmatrix}$ **(ii)** (4, 2, 6) **(iii)** 11.2

**10 (ii)** $\overrightarrow{AB} = \begin{pmatrix}-1\\2\\1\end{pmatrix}$; $\overrightarrow{AC} = \begin{pmatrix}8\\-4\\1\end{pmatrix}$; in both cases the scalar product = 0 **(iii)** 132.9° **(iv)** 8.08

**11 (i) (a)** 5 **(b)** $\sqrt{89}$ **(ii)** 62.2°
**(iii)** 20.9 **(iv)** (4, 6, –3)

**12 (i)** PQ: $\mathbf{r} = \begin{pmatrix}2\\2\\4\end{pmatrix} + \lambda\begin{pmatrix}-1\\2\\2\end{pmatrix}$; XY: $\mathbf{r} = \begin{pmatrix}-2\\-2\\-3\end{pmatrix} + \mu\begin{pmatrix}1\\2\\3\end{pmatrix}$
**(iii)** Yes **(iv)** Yes (1, 4, 6)

**13 (i)** $\mathbf{r} = \begin{pmatrix}4\\1\\3\end{pmatrix} + \lambda\begin{pmatrix}2\\3\\5\end{pmatrix}$ **(ii)** (0, –5, –7)
**(iii)** Q: (3, –1, 4) $A_1$: (2, –3, 5)
**(iv)** Both 78.5° (3s.f.)

**14 (ii)** $\begin{pmatrix}2\\-1\\3\end{pmatrix}$ **(iii)** (10, –5, 15)

**(iv)** OA: $\mathbf{r} = \mu\begin{pmatrix}5\\-12\\16\end{pmatrix}$; AB: $\mathbf{r} = \begin{pmatrix}5\\-12\\16\end{pmatrix} + \nu\begin{pmatrix}1\\5\\1\end{pmatrix}$ **(v)** 69°

**15 (i)** $\mathbf{r} = \begin{pmatrix}8\\0\\-4\end{pmatrix} + \lambda\begin{pmatrix}2\\1\\-1\end{pmatrix}$ **(ii)** (2, –3, –1)
**(iii)** $\sqrt{150}$ **(iv)** 3:2 **(v)** 8.48° (2d.p.)

**16 (i)** A(–50, –50, 0); B(50, –50, 0); C(50, 50, 0); D(–50, 50, 0); E(0, 0, 100)
**(ii)** ECD: $2y + z = 100$; EDA: $-2x + z = 100$; EAB: $-2y + z = 100$

**(iii)** $\mathbf{r} = \begin{pmatrix}0\\0\\20\end{pmatrix} + \lambda\begin{pmatrix}2\\0\\1\end{pmatrix}$, 35.8 m

**17 (i)** $\mathbf{r} = \begin{pmatrix} 2 \\ 0 \\ 15 \end{pmatrix} + \lambda\begin{pmatrix} 4 \\ 3 \\ -2 \end{pmatrix}$ **(ii)** $\begin{pmatrix} -1+4\lambda \\ 1+3\lambda \\ 14-2\lambda \end{pmatrix}$, (6, 3, 13)

**(iii)** 13 **(iv)** 1:2

**18 (i)** $\begin{pmatrix} 2 \\ -3 \\ 4 \end{pmatrix}$ **(ii)** $\mathbf{r} = \begin{pmatrix} 3 \\ -8 \\ 12 \end{pmatrix} + \lambda\begin{pmatrix} 2 \\ -3 \\ 4 \end{pmatrix}$; (−1, −2, 4)

**(iii)** (0, −3.5, 6) **(iv)** 15.6° (1d.p.)

**19 (i)** $\mathbf{r} = (2\mathbf{i} + 3\mathbf{j} + 5\mathbf{k}) + \lambda(3\mathbf{i} + \mathbf{j} - 2\mathbf{k})$

**(ii)** $\lambda = 1$; (5, 4, 3)

**(iii)** (9.5, 5.5, 0)

**(iv)** (6.5, 4.5, 2); 1.87 (3s.f.)

**(v)** $\mathbf{i} + 2\mathbf{j} = 3\mathbf{k}$; 38.2° (1d.p.)

**20 (i)** $2x - 3y + 72 = -5$

**(ii)** $\mathbf{r} = (130\mathbf{i} - 40\mathbf{j} + 20\mathbf{k}) + \lambda(8\mathbf{i} - 4\mathbf{j} + \mathbf{k})$

**(iii)** $10\mathbf{i} + 20\mathbf{j} + 5\mathbf{k}$

**(iv)** 135 m

**21 (i)** $\begin{pmatrix} a \\ b \\ 1 \end{pmatrix}$ **(ii)** $\overrightarrow{AB} = \begin{pmatrix} 2 \\ -3 \\ 0 \end{pmatrix}$; $\overrightarrow{AC} = \begin{pmatrix} 3 \\ -5 \\ 1 \end{pmatrix}$

**(iii)** $2a - 3b = 0$; $3a - 5b + 1 = 0$

**(iv)** $3x + 2y + z = 6$

**(v)** 36.7° (1d.p.)

**(vi)** $(3\frac{2}{3}, -3\frac{2}{3}, 2\frac{1}{3})$

**22 (i)** (6, 4.5, 3)

**(iii)** $x - 2z = 0$

**(iv)** AOBC: $\begin{pmatrix} 0 \\ 2 \\ -3 \end{pmatrix}$; DOBE: $\begin{pmatrix} 1 \\ 0 \\ -2 \end{pmatrix}$; 41.9° (1d.p.); 138.1°

**23 (i)** $\overrightarrow{AB} = 14\mathbf{i} + 2\mathbf{j} + 5\mathbf{k}$; $\overrightarrow{AD} = -5\mathbf{i} + 10\mathbf{j} + 10\mathbf{k}$; (13, 14, 18)

**(iii)** $\lambda = \frac{1}{3}, \mu = \frac{2}{15}$. It contains the origin

**(iv)** $2x + 11y - 10z = 0$

**24 (i)** $\mathbf{r} = \begin{pmatrix} 2 \\ 3 \\ 5 \end{pmatrix} + \lambda\begin{pmatrix} 1 \\ 1 \\ -0.5 \end{pmatrix}$

**(ii)** (12, 13, 0)

**(iii)** 109.5° (1d.p.)

**(iv)** 25 m

## Chapter 6

### Exercise 6A (Page 183)

**2** $\frac{ds}{dt} = \frac{k}{s^2}$

**3** $\frac{dh}{dt} = k\ln(H - h)$

**4** $\frac{dm}{dt} = \frac{k}{m}$

**5** $\frac{dP}{dt} = k\sqrt{P}$

**6** $\frac{de}{d\theta} = k\theta$

**7** $\frac{d\theta}{dt} = -\frac{(\theta - 15)}{160}$

**8** $\frac{dN}{dt} = \frac{N}{20}$

**9** $\frac{dv}{dt} = \frac{4}{\sqrt{v}}$

**10** $\frac{dA}{dt} = \frac{2k\sqrt{\pi}}{\sqrt{A}} = \frac{k'}{\sqrt{A}}$

**11** $\frac{d\theta}{ds} = -\frac{s}{4}$

**12** $\frac{dh}{dt} = -\frac{1}{8\pi h^2}$

**13** $\frac{dV}{dt} = -\frac{2V}{1125\pi}$

**14** $\frac{dh}{dt} = \frac{(2 - k\sqrt{h})}{100}$

### Exercise 6B (Page 188)

**1 (i)** $y = \frac{1}{3}x^3 + c$ **(ii)** $y = \sin x + c$ **(iii)** $y = e^x + c$

**(iv)** $y = \frac{2}{3}x^{\frac{3}{2}} + c$

**2 (i)** $y = -\frac{2}{(x^2 + c)}$ **(ii)** $y^2 = \frac{2}{3}x^3 + c$ **(iii)** $y = Ae^x$

**(iv)** $y = \ln|e^x + c|$ **(v)** $y = Ax$ **(vi)** $y = (\frac{1}{4}x^2 + c)^2$

**(vii)** $y = -\frac{1}{(\sin x + c)}$ **(viii)** $y^2 = A(x^2 + 1) - 1$

**(ix)** $y = -\ln(c - \frac{1}{2}x^2)$ **(x)** $y^3 = \frac{3}{2}x^2\ln x - \frac{3}{4}x^2 + c$

### Exercise 6C (Page 192)

**1 (i)** $y = \frac{1}{3}x^3 - x - 4$ **(ii)** $y = e^{x^3/3}$ **(iii)** $y = \ln(\frac{1}{2}x^2 + 1)$

**(iv)** $y = \frac{1}{(2 - x)}$ **(v)** $y = e^{(x^2 - 1)/2} - 1$ **(vi)** $y = \sec x$

**2 (i)** $\theta = 20 - Ae^{-2t}$ **(ii)** $\theta = 20 - 15e^{-2t}$

**(iii)** $t = 1.01$ seconds

**3 (i)** $N = Ae^t$ **(ii)** $N = 10e^t$

**(iii)** $N$ tends to $\infty$, which would never be realised because of the combined effects of food shortage, predators and human controls

**4** $s = \sqrt{4t + c}$

**5 (i)** $\frac{1}{3y} + \frac{1}{3(3 - y)}$ **(ii)** $\frac{1}{3}\ln\left|\frac{y}{3 - y}\right| + c$ or $\frac{1}{3}\ln\left|\frac{Ay}{3 - y}\right|$

**(iii)** $y = \frac{3x^3}{(4 + x^3)}$

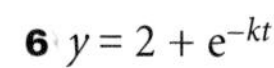

**6** $y = 2 + e^{-kt}$

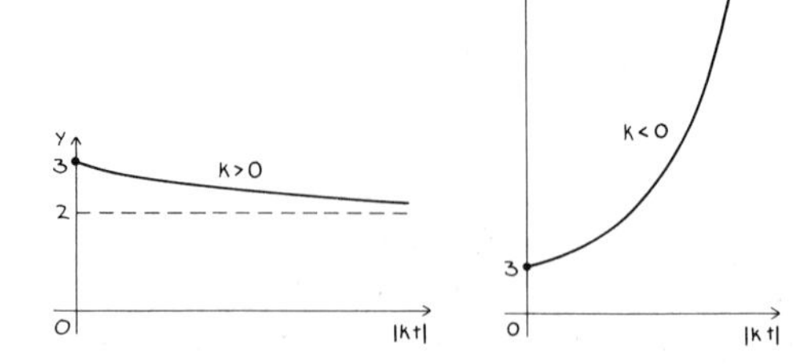


**7 (i)** $N = 1500e^{0.0347t} = 1500 \times 2^{t/20}$ **(ii)** $N = 24\,000$

**(iii)** Time taken = 11 hours 42 minutes

**8 (ii)** $\dfrac{1}{x-1} - \dfrac{1}{x+1}$ **(iii)** $y = \dfrac{(x+1)}{2(x-1)}e^{3-x}$

**10** $\dfrac{dx}{dt} = \dfrac{(2-3x)}{100}$, $x = \frac{1}{3}(2 - 2e^{-3t/100})$,

time taken = 40.8 seconds

**11 (i)** $\dfrac{dr}{dt} = \dfrac{k}{r^2}$ **(ii)** $k = 5000$; 141 m (3s.f.)

**(iii)** $\dfrac{dr}{dt} = \dfrac{k_1}{r^2(2+t)}$; $k_1 = 10\,000$ **(iv)** 104 m (3s.f.)

**12 (i)** $P = 600e^{kt}$

**(iii)** $P = 600e^{(0.005t - 0.4\sin(0.02t))}$; very good fit

**(iv)** 549

**13 (i)** $\dfrac{1}{3(2-x)} + \dfrac{1}{3(1+x)}$

**(ii)** $\frac{1}{3}$

**(iv)** 1.18 h (2d.p.)

**(v)** 0.728 kg

**14 (i)** $2x\sin 2x + \cos 2x + c$

**(iii)** $y^2 = 4x^2 + 4x\sin 2x + 2\cos 2x + 1$

**15 (i)** $y^2 = 4x$ **(ii)** $\dfrac{dy}{dx} = \dfrac{1}{t}$

**(iv)** $y^2 = -2x^2 + c$; $y^2 + 2x^2 = 4$

**(v)**

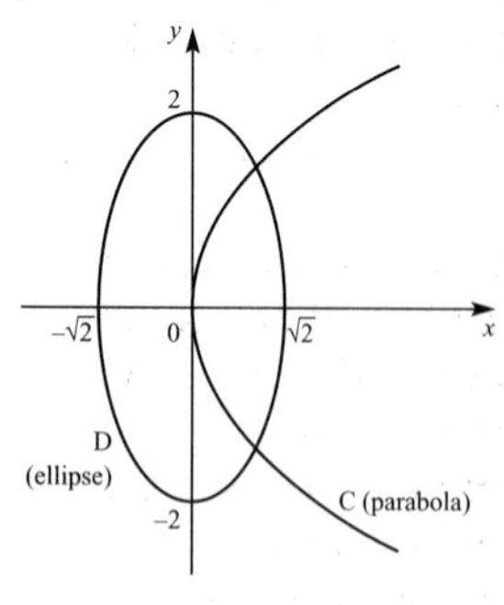


**16 (i)** $\dfrac{3}{(3x-1)} - \dfrac{1}{x}$

**(iii)** $t = 1.967$ (3d.p.)

**(iv)** 500 and 3550

**17 (ii)** $\cot x$; $\ln(\sin x) + c$

**(iii)** $y = 0.185$ (3s.f.); minimum

**18 (a) (i)** $\dfrac{dh}{ds} = ks$, $h = \frac{1}{2}ks^2$

**(ii)**

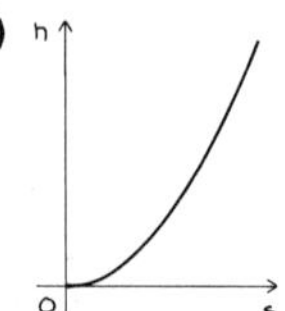


**(iii)** Unrealistic as wave height increases without limit, ever faster

**(b) (i)** $\dfrac{dh}{ds} = \dfrac{k}{s+5}$, $h = k\ln\left(\dfrac{s+5}{5}\right)$

**(ii)**

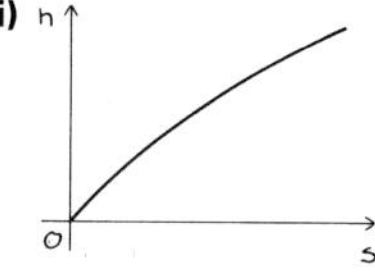


**(iii)** More realistic but still no limit to wave height

**(c) (i)** $\dfrac{dh}{ds} = ke^{-cs}$, $h = A(1 - e^{-cs})$

**(ii)**

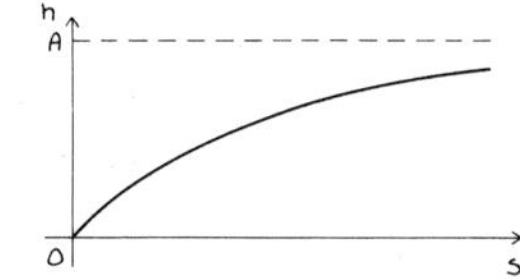


**(iii)** The most realistic

# Index

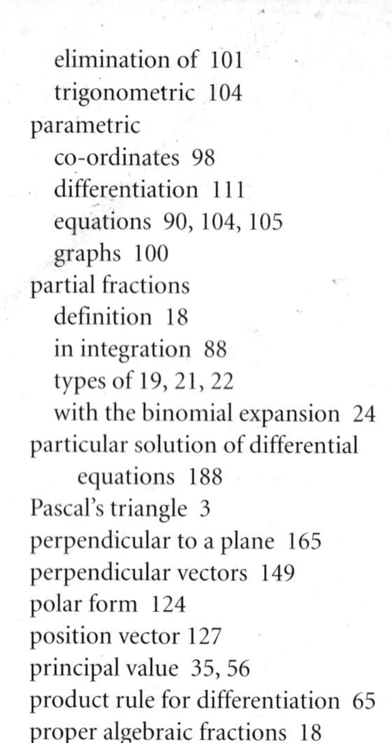